Modeling Crop Photosynthesis—from Biochemistry to Canopy

Modeling Crop Photosynthesis—from Biochemistry to Canopy

Proceedings of a symposium sponsored by Division C-2 of the Crop Science Society of America and Division A-3 of the American Society of Agronomy in Anaheim, California, 29 Nov. 1988.

Editors

K. J. Boote and R. S. Loomis

Organizing Committee

K. J. Boote

Editor-in-Chief CSSA

C. W. Stuber

Editor-in-Chief ASA

G. A. Peterson

Managing Editor

S. H. Mickelson

Assistant Editor

P. Kasper

CSSA Special Publication Number 19

Crop Science Society of America, Inc.
American Society of Agronomy, Inc.
Madison, Wisconsin, USA
1991

Crop Science Society of America, Inc.
American Society of Agronomy, Inc.
677 South Segoe Road, Madison, WI 53711, USA

Library of Congress Cataloging-in-Publication Data

Modeling crop photosynthesis—from biochemistry to canopy : proceedings of a symposium / sponsored by Division C-2 of the Crop Science Society of America and Division A-3 of the American Society of Agronomy in Anaheim, California, 29 Nov. 1988 ; editors, K.J. Boote and R.S. Loomis.
p. cm. — (CSSA special publication : no. 19)
Includes bibliographical references.
ISBN 0-89118-533-X
1. Photosynthesis—Computer simulation—Congresses. 2. Crops—Physiology—Computer simulation—Congresses. I. Boote, K.J. II. Loomis, R.S. III. Crop Science Society of America. Division C-2. IV. American Society of Agronomy. Division A-3. V. Series.
QK882.M8 1991
581.1'3342—dc20 91-26613
CIP

Printed in the United States of America

CONTENTS

FOREWORD

Photosynthesis is the most important biochemical process in nature. The rates at which plants fix CO_2, under the wide range of conditions found in nature, determine their productivity and ultimate utility to humans. As scientists have learned more about factors that limit and control this complex process, new genetic and management strategies have been devised to more fully exploit and develop its potential.

The models described in the chapters in this volume summarize the state of knowledge regarding the interactions of environmental and biochemical factors on crop photosynthesis. These models are not only valuable research tools but, when integrated into sophisticated crop simulation models, they become powerful tools to improve the management of crop production systems. Crop producers must strive to address environmental concerns and remain economically competitive. Computer models can be immensely valuable aids.

Crop simulation models need to accurately reflect actual biological happenings under complex and rapidly changing conditions. Such models must be based on accurate and reliable information regarding the most fundamental of plant processes — photosynthesis. The authors of this volume have made significant strides toward this end. Their work is to be commended.

V. L. LECHTENBERG, *president*
Crop Science Society of America

D. R. NIELSEN, *president*
American Society of Agronomy

PREFACE

Explanation and prediction of the growth of managed and natural ecosystems in response to climatic and soil-related factors are increasingly important as objectives of our science. Quantitative prediction of complex systems, however, depends on integrating information through levels of organization, and the principal approach we have for that is through the construction of simulation models. Simulation of the system's use and balance of C, beginning with the input of C from canopy assimilation, forms the essential core of most simulation models that deal with the growth of vegetation.

It is now more than 40 yr since the first detailed model of canopy light interception was reported by Monsi and Saeki (1948). That work was followed by more advanced geometrical and physiological models of foliage canopies by de Wit (1965), the Estonian group (summarized by Ross, 1981), and Duncan et al. (1967). The sophistication of the models for canopy light interception was not matched in the area of models for biochemistry–physiology of photosynthesis until the past 10 yr, when the Glasshouse Crops Group in England (Acock, Ch. 3 in this book) and Farquhar and von Caemmerer (1982) began to publish advanced approaches to that problem.

Parallel progress was made in the development of dynamic simulation models of the growth of crops. Crop model development was stimulated by Forrester's (1961) state-variable approach to system simulation and by access to mainframe computers and, more recently, powerful microcomputers. Of necessity, early crop models employed simple, summary approaches to the simulation of photosynthesis. Presently, simulation studies are no longer limited by one's mainframe budgets, and quite complicated models are processed quickly on microcomputers. It is now appropriate to determine whether more sophisticated approaches for light interception, photosynthetic biochemistry, and canopy photosynthesis can be incorporated into crop models, and to determine the utility of adding such sophisticated approaches in contrast to refining summary approaches.

In this publication, the contributing authors have summarized some of the approaches now used to predict leaf and canopy photosynthesis. Most of these models can stand alone for studies of photosynthesis, or they can be incorporated into crop growth models. Models for single-leaf response to light, CO_2, and temperature are succinctly described in the first two chapters (Evans and Farquhar, Ch. 1; Harley and Tenhunen, Ch. 2). Norman and Arkebauer (Ch. 5) and Gutschick (Ch. 4) illustrate how numerical, layered-canopy, simulation models describing complete radiation, energy, water vapor, and CO_2 balances among leaf strata can be used to predict whole-canopy assimilation response to light, CO_2, wind speed, humidity, and temperature. Gutschick shows how it is possible to reduce a complex numerical model into a summary model that provides important insights for agricultural production and water-use efficiency. Acock (Ch. 3) introduces alternative approaches for predicting whole-canopy response to light, CO_2, and temperature. Sinclair's chapter (Ch. 6) contributes an important advance in simple canopy models, particularly incorporation of the effects of leaf N content on canopy assimilation. Lastly, Boote and Loomis (Ch. 7) review the approaches taken by the various authors to describe leaf and canopy assimilation processes, and then present simplified equations for predicting canopy assimilation response to light, leaf area index, and incomplete hedgerow canopy coverage.

REFERENCES

de Wit, C.T. 1965. Photosynthesis of leaf canopies. Agric. Res. Rep. no. 663. PUDOC, Wageningen, the Netherlands.

Duncan, W.G., R.S. Loomis, W.A. Williams, and R. Hanau. 1967. A model for simulating photosynthesis in plant communities. Hilgardia 38:181–205.

Farquhar, G.D., and S. von Caemmerer. 1982. Modeling of photosynthetic response to environment. p. 549–587. *In* O.L. Lang et al. (ed.) Physiological plant ecology II. New Ser. Vol. 12B. Encyclopedia of plant physiology. Springer-Verlag, Berlin.

Forrester, J.W. 1961. Industrial dynamics. Massachusetts Inst. Technol. Press, Cambridge, MA.

Monsi, M., and T. Saeki. 1953. Über den Lickhfaktor in den Pflangengesellschaften und seine Bedeutung für die Stoffproduktion. Jpn. J. Bot. 14:22–52.

Ross, J. 1981. The radiation regime and architecture of plant stands. Tasks for vegetative sciences no. 3. Junk, The Hague.

K.J. BOOTE and R.S. LOOMIS
Editors

CONTRIBUTORS

Basil Acock	Research Leader, USDA-ARS, NRI, Systems Research Laboratory, BARC-West, Beltsville, MD 20705-2350
T. J. Arkebauer	Assistant Professor, Department of Agronomy, University of Nebraska, Lincoln, NE 68583-0817
K. J. Boote	Professor of Agronomy, Agronomy Department, University of Florida, Gainesville, FL 32611
John R. Evans	Professor, P.E.B. Research School of Biological Sciences, Australian National University, Canberra 2601, Australia
Graham D. Farquhar	Professor, P.E.B. Research School of Biological Sciences, Australian National University, Canberra 2601, Australia
Vincent P. Gutschick	Professor of Biology, Department of Biology, New Mexico State University, Las Cruces, NM 88003
Peter C. Harley	Research Scientist, Systems Ecology Research Group, San Diego State University, San Diego, CA 92182
R. S. Loomis	Professor of Agronomy, Department of Agronomy and Range Science, University of California, Davis, CA 95616
J. M. Norman	Professor of Soil Science, Department of Soil Science, University of Wisconsin, Madison, WI 53706
Thomas R. Sinclair	Plant Physiologist, USDA-ARS, Agronomy-Physiology Laboratory, University of Florida, Gainesville, FL 32611
J. D. Tenhunen	Associate Director, Systems Ecology Research Group, San Diego State University, San Diego, CA 92182

Conversion Factors for SI and non-SI Units

Conversion Factors for SI and non-SI Units

To convert Column 1 into Column 2, multiply by	Column 1 SI Unit	Column 2 non-SI Unit	To convert Column 2 into Column 1, multiply by
		Length	
0.621	kilometer, km (10^3 m)	mile, mi	1.609
1.094	meter, m	yard, yd	0.914
3.28	meter, m	foot, ft	0.304
1.0	micrometer, μm (10^{-6} m)	micron, μ	1.0
3.94×10^{-2}	millimeter, mm (10^{-3} m)	inch, in	25.4
10	nanometer, nm (10^{-9} m)	Angstrom, Å	0.1
		Area	
2.47	hectare, ha	acre	0.405
247	square kilometer, km^2 $(10^3\ m)^2$	acre	4.05×10^{-3}
0.386	square kilometer, km^2 $(10^3\ m)^2$	square mile, mi^2	2.590
2.47×10^{-4}	square meter, m^2	acre	4.05×10^3
10.76	square meter, m^2	square foot, ft^2	9.29×10^{-2}
1.55×10^{-3}	square millimeter, mm^2 $(10^{-3}\ m)^2$	square inch, in^2	645
		Volume	
9.73×10^{-3}	cubic meter, m^3	acre-inch	102.8
35.3	cubic meter, m^3	cubic foot, ft^3	2.83×10^{-2}
6.10×10^4	cubic meter, m^3	cubic inch, in^3	1.64×10^{-5}
2.84×10^{-2}	liter, L (10^{-3} m^3)	bushel, bu	35.24
1.057	liter, L (10^{-3} m^3)	quart (liquid), qt	0.946
3.53×10^{-2}	liter, L (10^{-3} m^3)	cubic foot, ft^3	28.3
0.265	liter, L (10^{-3} m^3)	gallon	3.78
33.78	liter, L (10^{-3} m^3)	ounce (fluid), oz	2.96×10^{-2}
2.11	liter, L (10^{-3} m^3)	pint (fluid), pt	0.473

	Mass		
2.20×10^{-3}	gram, g (10^{-3} kg)	pound, lb	454
3.52×10^{-2}	gram, g (10^{-3} kg)	ounce (avdp), oz	28.4
2.205	kilogram, kg	pound, lb	0.454
0.01	kilogram, kg	quintal (metric), q	100
1.10×10^{-3}	kilogram, kg	ton (2000 lb), ton	907
1.102	megagram, Mg (tonne)	ton (U.S.), ton	0.907
1.102	tonne, t	ton (U.S.), ton	0.907
	Yield and Rate		
0.893	kilogram per hectare, kg ha^{-1}	pound per acre, lb $acre^{-1}$	1.12
7.77×10^{-2}	kilogram per cubic meter, kg m^{-3}	pound per bushel, bu^{-1}	12.87
1.49×10^{-2}	kilogram per hectare, kg ha^{-1}	bushel per acre, 60 lb	67.19
1.59×10^{-2}	kilogram per hectare, kg ha^{-1}	bushel per acre, 56 lb	62.71
1.86×10^{-2}	kilogram per hectare, kg ha^{-1}	bushel per acre, 48 lb	53.75
0.107	liter per hectare, L ha^{-1}	gallon per acre	9.35
893	tonnes per hectare, t ha^{-1}	pound per acre, lb $acre^{-1}$	1.12×10^{-3}
893	megagram per hectare, Mg ha^{-1}	pound per acre, lb $acre^{-1}$	1.12×10^{-3}
0.446	megagram per hectare, Mg ha^{-1}	ton (2000 lb) per acre, ton $acre^{-1}$	2.24
2.24	meter per second, m s^{-1}	mile per hour	0.447
	Specific Surface		
10	square meter per kilogram, m^2 kg^{-1}	square centimeter per gram, cm^2 g^{-1}	0.1
1000	square meter per kilogram, m^2 kg^{-1}	square millimeter per gram, mm^2 g^{-1}	0.001
	Pressure		
9.90	megapascal, MPa (10^6 Pa)	atmosphere	0.101
10	megapascal, MPa (10^6 Pa)	bar	0.1
1.00	megagram per cubic meter, Mg m^{-3}	gram per cubic centimeter, g cm^{-3}	1.00
2.09×10^{-2}	pascal, Pa	pound per square foot, lb ft^{-2}	47.9
1.45×10^{-4}	pascal, Pa	pound per square inch, lb in^{-2}	6.90×10^3

(continued on next page)

Conversion Factors for SI and non-SI Units

To convert Column 1 into Column 2, multiply by	Column 1 SI Unit	Column 2 non-SI Unit	To convert Column 2 into Column 1, multiply by
		Temperature	
1.00 (K − 273)	Kelvin, K	Celsius, °C	1.00 (°C + 273)
(9/5 °C) + 32	Celsius, °C	Fahrenheit, °F	5/9 (°F − 32)
		Energy, Work, Quantity of Heat	
9.52×10^{-4}	joule, J	British thermal unit, Btu	1.05×10^{3}
0.239	joule, J	calorie, cal	4.19
10^{7}	joule, J	erg	10^{-7}
0.735	joule, J	foot-pound	1.36
2.387×10^{-5}	joule per square meter, J m^{-2}	calorie per square centimeter (langley)	4.19×10^{4}
10^{5}	newton, N	dyne	10^{-5}
1.43×10^{-3}	watt per square meter, W m^{-2}	calorie per square centimeter minute (irradiance), cal cm^{-2} min^{-1}	698
		Transpiration and Photosynthesis	
3.60×10^{-2}	milligram per square meter second, mg m^{-2} s^{-1}	gram per square decimeter hour, g dm^{-2} h^{-1}	27.8
5.56×10^{-3}	milligram (H_2O) per square meter second, mg m^{-2} s^{-1}	micromole (H_2O) per square centimeter second, μmol cm^{-2} s^{-1}	180
10^{-4}	milligram per square meter second, mg m^{-2} s^{-1}	milligram per square centimeter second, mg cm^{-2} s^{-1}	10^{4}
35.97	milligram per square meter second, mg m^{-2} s^{-1}	milligram per square decimeter hour, mg dm^{-2} h^{-1}	2.78×10^{-2}
		Plane Angle	
57.3	radian, rad	degrees (angle), °	1.75×10^{-2}

	Electrical Conductivity, Electricity, and Magnetism		
10	siemen per meter, S m^{-1}	millimho per centimeter, mmho cm^{-1}	0.1
10^4	tesla, T	gauss, G	10^{-4}
	Water Measurement		
9.73×10^{-3}	cubic meter, m^3	acre-inches, acre-in	102.8
9.81×10^{-3}	cubic meter per hour, $m^3 h^{-1}$	cubic feet per second, $ft^3 s^{-1}$	101.9
4.40	cubic meter per hour, $m^3 h^{-1}$	U.S. gallons per minute, gal min^{-1}	0.227
8.11	hectare-meters, ha-m	acre-feet, acre-ft	0.123
97.28	hectare-meters, ha-m	acre-inches, acre-in	1.03×10^{-2}
8.1×10^{-2}	hectare-centimeters, ha-cm	acre-feet, acre-ft	12.33
	Concentrations		
1	centimole per kilogram, cmol kg^{-1} (ion exchange capacity)	milliequivalents per 100 grams, meq 100 g^{-1}	1
0.1	gram per kilogram, g kg^{-1}	percent, %	10
1	milligram per kilogram, mg kg^{-1}	parts per million, ppm	1
	Radioactivity		
2.7×10^{-11}	becquerel, Bq	curie, Ci	3.7×10^{10}
2.7×10^{-2}	becquerel per kilogram, Bq kg^{-1}	picocurie per gram, pCi g^{-1}	37
100	gray, Gy (absorbed dose)	rad, rd	0.01
100	sievert, Sv (equivalent dose)	rem (roentgen equivalent man)	0.01
	Plant Nutrient Conversion		
	Elemental	***Oxide***	
2.29	P	P_2O_5	0.437
1.20	K	K_2O	0.830
1.39	Ca	CaO	0.715
1.66	Mg	MgO	0.602

1 Modeling Canopy Photosynthesis from the Biochemistry of the C_3 Chloroplast

John R. Evans and Graham D. Farquhar

Australian National University
Canberra, Australia

Photosynthesis involves the interception of light energy and its conversion to chemical energy in intermediates of high chemical potential, which are then used to drive the catalytic fixation of CO_2 into sugars and other compounds. Hundreds of different proteins are involved along the way but, despite this complexity, there are several key factors that allow simplification in our model of the system. Attention can be focused on the principal CO_2-fixing enzyme ribulose-1,5-bisphosphate (RuBP) carboxylase-oxygenase, Rubisco, which is the most abundant leaf protein. To achieve adequate rates of CO_2 assimilation, Rubisco needs to be abundant because it has a low affinity for CO_2 and a relatively slow rate of catalysis. Rubisco also catalyzes the competitive reaction between RuBP and O_2 and considerable metabolic effort by the cell is required to recover the C skeleton in phosphoglycolate that is so produced. The kinetics of the Rubisco enzyme, with respect to its substrates RuBP, CO_2, and O_2, encompass a large proportion of the photosynthetic properties of a leaf. We will present the basic equations that have been discussed in detail elsewhere (Farquhar & von Caemmerer, 1982), with particular emphasis on how they apply to canopy photosynthesis. Since the publication of the model by Farquhar et al. (1980), considerable experimental evidence has been obtained that substantiates much of the theory and has enriched the detail of the underlying biochemical mechanisms of the model.

In order to gain CO_2, the leaf loses water to the atmosphere. The amount of water lost per C gained depends, firstly, on the water vapor-pressure difference between the leaf and the air. Second, it depends on the intercellular partial pressure of CO_2, pCO_2. Conventional methods of assessing the transpiration-use efficiency, W (amount of C gained per water used), involve careful measurements of soil moisture by either weighing pots or using neutron probes. This has proved rather impractical on the scale necessary for plant breeding programs. A new technique that involves the determination of the $^{13}C/^{12}C$ ratio of the plant can be used to assess the integrated value of intercellular pCO_2 and this enables transpiration-use efficiency to become a selection criterion. The underlying theory (Farquhar et al., 1982) is currently being evaluated in the context of plant breeding with several crops (e.g., wheat [*Triticum aestivum* L.], Farquhar & Richards, 1984; Condon et al., 1987; and peanut [*Arachis hypogaea* L.], Hubick et al., 1986. See also Farquhar et al., 1988). The basic equations will be presented here because this technique offers an exciting new avenue for plant improvement where yield is limited by the availability of water.

RATE OF CARBON DIOXIDE ASSIMILATION

The absorption of light by the pigments in the chloroplast membranes leads to the transfer of electrons from H_2O to nicotinamide adenine dinucleotide phosphate ($NADP^+$) to make NADPH, and the buildup of protons in the lumen of the thylakoids. The protons drive the regeneration of the high-energy compound adenosine triphosphate (ATP), catalyzed by the coupling factor. These two high-potential intermediates, NADPH and ATP, are used in the reactions of the C-reduction cycle to regenerate the substrate for CO_2 fixation, RuBP. Because of the high concentration of Rubisco in the chloroplast, the kinetics of the enzyme with respect to its substrate RuBP do not follow normal Michaelis-Menten kinetics (Farquhar, 1979). Rather, we think of Rubisco as either being limited by RuBP or not. The pool size of RuBP is small, and without continuous regeneration would be consumed within seconds. Since RuBP regeneration is closely coupled to the rate of electron transport and photophosphorylation, the RuBP-limited Rubisco velocity closely reflects the rate of electron transport. When RuBP is saturating, the photosynthetic properties reflect the affinity of Rubisco for CO_2 and the relative rates of oxygenation and carboxylation. The potential rate of electron transport declines at lower temperatures to a greater extent than does Rubisco activity; thus, the balance between the two capacities changes with temperature. In some low-temperature situations, phosphate recycling to the chloroplast prevents the potential rate of electron transport from being reached (Sharkey, 1985; Sage & Sharkey, 1987; Labate & Leegood, 1988). We will focus on the electron-transport properties, because the photosynthetic rate of many crop canopies is primarily light limited.

IRRADIANCE RESPONSE CURVES

The many steps between light absorption and RuBP regeneration, in combination with the complexity of the optics of the leaf, mean that a precise theoretical justification for Eq. [1] is not possible at present. However, the following equation can describe very precisely the relationship between potential electron transport rate, J, and the irradiance usefully absorbed by Photosystem II, I_2:

$$\Theta J^2 - (I_2 + J_{max})J + I_2 J_{max} = 0 \quad [1]$$

which can be solved for J as follows:

$$J = \{I_2 + J_{max} - [(I_2 + J_{max})^2 - 4\Theta I_2 J_{max}]^{1/2}\}/2\Theta \qquad [2]$$

where I_2 is related to the incident irradiance (400–700 nm), I_0, as follows: $I_2 = I_0 (1 - f)(1 - r)/2$. The factor f corrects for the spectral imbalance of the light (~0.15, see Evans, 1987a), r is the reflectance plus any small transmittance of the leaf or crop to photosynthetically active radiation (~0.12); I_0 is divided by 2 because light is absorbed by both Photosystem II and Photosystem I to drive one electron from H_2O to $NADP^+$. The maximum rate of electron transport, J_{max}, is a property of the thylakoids that varies depending on growth conditions. The factor Θ is a curvature factor, $0 \leq \Theta \leq 1$, which determines how quickly the transition is made from the region of maximum quantum yield to the light-saturated rate. When $\Theta = 0$, the equation degenerates to a rectangular hyperbola, while $\Theta = 1$ describes the Blackman response of two straight lines representing light-dependent and light-saturated rates.

The region of maximum quantum yield is found at low irradiance, where the rate of photosynthesis is linearly related to the irradiance. No significant variation is seen across a broad range of C_3 plants in the quantum yield measured as O_2 evolution in saturating CO_2, when expressed on an absorbed-light basis (Björkman & Demmig, 1987; Evans, 1987a). The absolute value of the quantum yield depends on the wavelength or spectral composition of the light (McCree, 1972; Inada, 1976). For sunlight, the quantum yield is about 15% below the maximum, which occurs with 600-nm light (Evans, 1987a). To correct for this, the incident irradiance is multiplied by the term $(1 - f)$.

The light-saturated rate of electron transport per unit leaf area is determined primarily by two factors. First, the capacity scales in proportion with the chlorophyll content per unit leaf area (Fig. 1–1A). This reflects the amount of photosynthetic apparatus in a given leaf area. Leaves that develop with a restricted N supply contain less N per unit leaf area. This corresponds to smaller protein contents in all fractions of the leaf. A similar situation can be reached during senescence, where N is progressively remobilized from the leaf.

The second determinant of the electron-transport capacity relates to the irradiance during growth of the leaf. The electron-transport capacity per unit of chlorophyll is less in leaves acclimated to low irradiance (Fig. 1–1B). This reflects the altered composition of the thylakoid membranes. When acclimated to low irradiance, thylakoid membranes are enriched in the light-harvesting chlorophyll a/b protein complex and depleted in Photosystem II reaction-center complexes, plastoquinone, cytochrome b/f complexes, coupling factor, and ferredoxin NADP reductase (Anderson, 1986). The electron-transport capacity increases as the relative abundance of plastoquinone, cytochrome f, coupling factor, and ferredoxin NADP reductase increases (Evans, 1987b; Terashima & Evans, 1988). The value of J_{max} correlates strongly with the cytochrome f content of the leaf (Terashima &

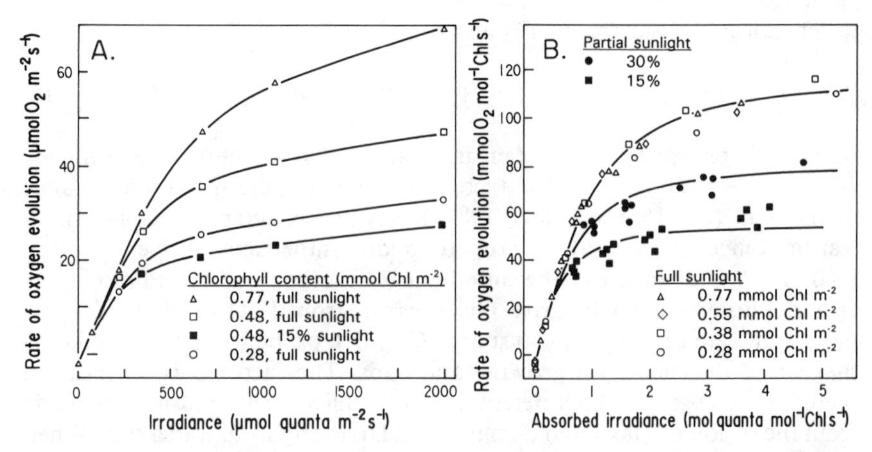

Fig. 1–1. Irradiance response curves of spinach leaves measured in a leaf-disk, O_2 electrode (Delieu & Walker, 1981) at 25° C and 1% CO_2, expressed on the basis of (A) leaf area or (B) chlorophyll content. Plants were grown under full ($\approx$2 mmol quanta m^{-2} s^{-1}) or partial sunlight with different NO_3^- nutrition, which caused the leaf chlorophyll content to vary. Lines were calculated using Eq. [2], with J_{max} = 500, 210, and 143 mmol e^- (mol Chl)$^{-1}$ s^{-1} for the 100, 30, and 15% irradiance treatments, respectively, Θ = 0.69 and f = 0.15. (Data from Evans & Terashima, 1987; Terashima & Evans, 1988).

Evans, 1988; Evans, 1988). Two of the curves in Fig. 1–1A are from leaves with the same chlorophyll content (0.48 mmol Chl m^{-2}), but the leaf for the lower curve was grown at 15% of full sunlight. The effect of N content can be separated from the effect of growth irradiance by expressing both axes on the basis of chlorophyll rather than leaf area (Fig. 1–1B). The upper curve represents leaves grown at 100% sunlight and, although their chlorophyll contents differed by a factor of 2.7, the same curve describes them all. The lower two curves represent leaves grown at 30 and 15% of full sunlight. They have the same quantum yield but lower electron-transport capacities per unit of chlorophyll.

The values for J_{max} calculated from gas-exchange characteristics have been compared with the corresponding in vitro uncoupled, light-saturated electron-transport activities for leaves of common bean (*Phaseolus vulgaris* L.) (von Caemmerer & Farquhar, 1981) and spinach (*Spinacia oleracea* L.) (Evans & Terashima, 1988). Variation in J_{max} was obtained by varying the N nutrition or irradiance during growth. In both species, good correlations were found, although the in vitro electron-transport activities were too small to account for the calculated J_{max}. This may simply reflect the incomplete extraction of leaves, damage to the thylakoids, or suboptimal assay conditions. It is frequently observed that Rubisco activity is also barely sufficient to account for the observed rates of CO_2 assimilation by leaves. As discussed below (Fig. 1–2A), the electron-transport and Rubisco capacities co-vary such that, for leaves at high irradiance and at ambient pCO_2, electron-transport and Rubisco capacities co-limit the rate of CO_2 assimilation.

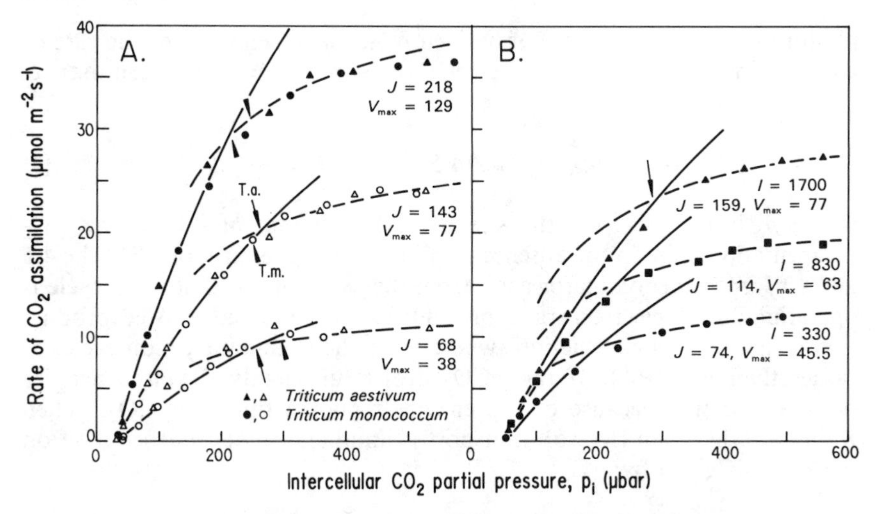

Fig. 1-2. Assimilation rate vs. intercellular partial pressure of CO_2 for (A) leaves grown with 12 m*M* NO_3^- (closed symbols) or 0.5 m*M* NO_3^- (open symbols), measured at 180 μmol quanta $m^{-2} s^{-1}$, 23 ° C, 13 mbar leaf-to-air vapor pressure difference, and 200 mbar pO_2 (data from Evans, 1985); and (B) a leaf of *T. aestivum* measured at irradiances (I) of 1700, 830, and 330 μmol quanta $m^{-2} s^{-1}$ (data from Evans, 1986). Model lines are generated with $\Gamma_* = 37$ μbar, $K_c = 252$ μbar, $K_o = 192$ mbar, $O = 210$ mbar, and $R_d = 1$ μmol $m^{-2} s^{-1}$. Arrows indicate the operating point with 340 μbar pCO_2 external to the leaf. Model lines are generated from Eq. [3] (broken lines) and [4] (solid lines).

The third parameter in the irradiance response curve that needs discussion is the curvature factor Θ. When fitting the curves, Θ and the quantum yield interact such that low values for quantum yield are associated with Θ values approaching 1 (Leverenz, 1987). With high quantum yields corrected for spectral quality, Θ values of about 0.7 have been found for spinach leaves with widely varying chlorophyll contents and for spinach and pea (*Pisum sativum* L.) leaves acclimated to a wide range of irradiances (Fig. 1-1; Evans, 1987b; Evans & Terashima, 1987; Terashima & Evans, 1988). It is not known what determines the curvature factor. Acclimation of chloroplasts to the intraleaf light gradient has some influence. Smaller values of Θ are observed when the irradiance response curves are measured using unilateral illumination from the abaxial rather than the normal adaxial surface (Oja & Laisk, 1976; Terashima & Saeki, 1985; Terashima, 1986). The value of Θ was also seen to increase as conifer needles acclimated to the light environment inside an integrating sphere (Leverenz, 1988).

DEPENDENCE ON PARTIAL PRESSURE OF CARBON DIOXIDE

The rate of CO_2 assimilation can vary for a given rate of electron transport, depending on the relative partial pressures of CO_2 and O_2. At lower pCO_2, the rate of CO_2 assimilation decreases despite a constant rate of electron transport and RuBP regeneration, because the rate of photorespiration increases. For a given pO_2, the dependence of the rate of CO_2 assimilation, A, on the pCO_2, is modeled by (Farquhar & von Caemmerer, 1982):

$$A = J(p_c - \Gamma_*)/(4.5p_c + 10.5\Gamma_*) - R_d \quad [3]$$

where p_c is the pCO_2 at the site of carboxylation, and Γ_* is the CO_2 compensation point in the absence of mitochondrial respiration, R_d. The 4.5 and 10.5 in the denominator arise from the assumptions that no Q cycle is operating during electron transport and that the shortfall in ATP production is met by Mehler reaction (where O_2 is the terminal election acceptor rather than $NADP^+$). If the pCO_2 drops sufficiently, Rubisco activity becomes limiting because of the enzyme's poor affinity for CO_2. When Rubisco rather than electron transport is limiting, the dependence of A on the pCO_2 is given by:

$$A = V_{max}(p_c - \Gamma_*)/[p_c + K_c (1 + O/K_o)] - R_d \quad [4]$$

where V_{max} is the Rubisco activity with saturating CO_2, K_c and K_o are the Michaelis constants for CO_2 and O_2, respectively, and O is the pO_2.

As with electron-transport capacity, Rubisco capacity varies depending on the N content of the leaf and on the instantaneous irradiance. These effects are illustrated in Fig. 1–2. The response of the rate of CO_2 assimilation to the intercellular pCO_2 is shown for six leaves of *Triticum* (Fig. 1–2A). All the curves were measured with an irradiance of 1800 μmol quanta m^{-2} s^{-1} and the variation is due to different amounts of leaf N and, hence, different Rubisco and electron-transport capacities. The model lines were generated using Eq. [3] and [4] and show that, at high irradiance and normal pCO_2 (arrows), the rate of CO_2 assimilation falls close to the region of transition from Rubisco to electron-transport limitation. When the curves are measured at different irradiances, the rate of electron transport is directly affected (dashed lines, Fig. 1–2B). Rubisco activity also changes through the deactivation of catalytic sites (von Caemmerer & Edmondson, 1986). Deactivation occurs either from the loss of the CO_2–Mg^{2+} from the enzyme active site or due to the synthesis of a tight binding inhibitor such as 2-carboxy-arabinitol 1-phosphate (Sharkey et al., 1986a; Kobza & Seemann, 1988). Although Rubisco activity is reduced, it does not limit the rate of CO_2 assimilation to any greater extent than is imposed by the reduced electron-transport rate. Consequently, the simple form of Eq. [4] is generally sufficient.

EARLY CROP GROWTH

Having described the rate of CO_2 assimilation by leaves in terms of their biochemistry, we wish to demonstrate the link between photosynthesis and crop growth. Initially, crop growth is exponential as the seedlings rapidly expand their leaf area and intercept more light. As adjacent plants begin encroaching on one another, growth becomes linear. Classical growth analysis has defined relative growth rate as the product of net assimilation rate and leaf-area ratio. Net assimilation rate is, effectively, the assimilated CO_2 that is converted to dry matter. Leaf-area ratio is the area of leaf per unit plant dry matter and approximates the ability of the young plant to intercept light. We define net assimilation rate as the product of the rate of net CO_2 assimilation, A (mol C m^{-2} of leaf s^{-1}), the photoperiod as a fraction of the day, l (dimensionless), and one minus the fraction of net daily C that is respired, ϕ_c (dimensionless). Instead of leaf-area ratio, the reciprocal is used. The ratio ρ is the current ratio between total plant C and leaf area (mol C m^{-2} of leaf). The equation for relative growth rate, RGR (s^{-1}) is:

$$\mathrm{RGR} = \frac{1}{M}\frac{\mathrm{d}M}{\mathrm{d}t} = Al(1 - \phi_c)/\rho \qquad [5]$$

where M is the total plant C and dM/dt is the change in plant C per unit time (Masle & Farquhar, 1988). Equation [5] can describe growth in a controlled environment where the irradiance is constant throughout the photoperiod. When irradiance varies, it is necessary to integrate A through the day using Eq. [2] and [3], where I_0 is the intercepted irradiance.

The fraction of daily C that is respired, ϕ_c, is difficult to assess in field situations, but has been measured in controlled environments. The value of ϕ_c for use in Eq. [5] is defined as the fraction of daily *net* CO_2 assimilated that is respired:

$$\phi_c = [R_r^L\, L + (R_r^N + R_s^N)N]/(sAL) \qquad [6]$$

where R_r^L, R_r^N and R_s^N are the respiration rates of the root in the photoperiod and at night and that of the shoot at night, respectively, L and N are the duration of the photoperiod and night (s), s is the leaf area (m^2), and A is the net rate of CO_2 assimilation of the shoot expressed per unit leaf area. For wheat, values range from 0.30 to 0.35 (King & Evans, 1967; Evans, 1983; Masle & Farquhar, 1988) and other estimates range from 0.3 to 0.5 (Koh & Kumura, 1975; McCree, 1986).

SCALING FROM THE CHLOROPLAST TO THE CANOPY

The capacity for photosynthesis depends on the nutrient content and light environment of the leaf. It is now well documented that the light gradient within a bifacial leaf is accompanied by changes in the chloroplast properties (Terashima & Inoue, 1985). That is, near the illuminated surface, the chloroplasts have higher rates of electron transport and Rubisco activities per unit of chlorophyll than chloroplasts farther away from the surface. Despite the inhomogeneity across the leaf, the photosynthetic properties of the leaf can be described simply as the sum of all the chloroplast capacities in a given unit area if the chloroplast properties scale with the intraleaf light gradient (Terashima & Saeki, 1983, 1985; Terashima & Inoue, 1985; Farquhar, 1989).

The same argument applies when the properties of a canopy of leaves are examined. Both the N contents of and the irradiance perceived by leaves in a crop canopy will vary during the growth of a crop. Aging of the leaves will generally be associated with remobilization of N and a reduction in the chlorophyll content. The production of new leaves above will reduce the irradiance reaching lower leaves and, consequently, reduce the electron-transport capacity per unit of chlorophyll lower in the canopy. These two factors will create a gradient in electron-transport capacity of leaves down the canopy. Gradients in leaf N content with respect to the relative irradiance have been found in canopies of tall goldenrod (*Solidago altissima* L.) (Hirose & Werger, 1987), guar [*Cyamopsis tetragonoloba* (L.) Taubert] (Charles-Edwards et al., 1987), peach [*Prunus persica* (L.) Batsch var. *persica*] (DeJong & Doyle, 1985), and pepper (*Piper* spp.) (Walters & Field, 1987). If the capacity of any leaf reflects the irradiance it receives and Θ does not vary between leaves, then the canopy can be treated as a big leaf and the same equations apply.

In many unstressed leaf canopies, the photosynthetic rate will be limited by light for the majority of the day. Consequently, we can, to a first approximation, use Eq. [2] and [3] to describe light and CO_2 responses in leaf canopies. The rate of CO_2 assimilation was measured under a range of conditions for small wheat (*T. aestivum* and *T. monococcum* L.) crops (0.6 m^2) in growth cabinets. The J_{max} of the canopy was estimated by measuring irradiance response curves on several leaves and multiplying the average J_{max} per leaf by the leaf area index, LAI, of the canopy. The leaves were sampled to represent the vertical profile of the canopy. It was necessary to measure the respiration rate of the crop. Two examples of the rate of CO_2 assimilation by the canopy as a function of irradiance are shown in Fig. 1–3. For *T. aestivum*, the J_{max} per unit leaf area was slightly greater than for *T. monococcum* and the LAI was 7.1, compared with 3. Consequently, the J_{max} values for the canopies differed by nearly a factor of 3. Within the range of irradiances that it was possible to achieve in the cabinets, the relationships between the rate of CO_2 assimilation and irradiance were nearly linear. Even at full sunlight, little deviation from linearity was expected, which was not the case for single leaves (Fig. 1–1). The canopy with the higher J_{max} did have higher rates of CO_2 assimilation. However, the predicted ratio of A at 2000 μmol quanta m^{-2} s^{-1} was only 1.3, despite a 2.6-fold difference in J_{max} between the canopies. The agreement between the data and the curve predicted from single-leaf measurements was good in both cases.

For the CO_2 responses, wheat crops were grown with different levels of NO_3^- nutrition in a glasshouse before being transferred to the chamber for measurement at heading. As single-leaf measurements were not made in the experiment, J_{max} was calculated from the rate at ambient pCO_2 (p_a). The curve was then calculated using the electron-transport-limited Eq. [3] and it was possible to assume $p_c/p_a = 0.67$ rather than include a stomatal effect. It was also necessary to measure the whole-canopy dark respiration rate (Fig. 1–4). Again, the agreement between the data and the curve was excellent, considering the simplifying assymptions made. The canopy data

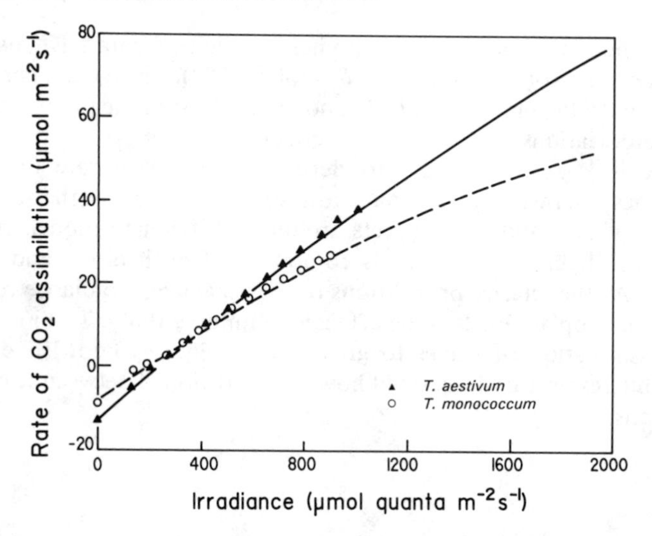

Fig. 1-3. Irradiance response curves for wheat crop canopies grown in a cabinet: *Triticum aestivum*: LAI = 7.1, J_{max} = 1450 μmol e^- m^{-2} s^{-1}, R_d = 12.3 μmol CO_2 m^{-2} s^{-1}; *T. monococcum*: LAI = 3.0; J_{max} = 550 μmol e^- m^{-2} s^{-1}, R_d = 9 μmol CO_2 m^{-2} s^{-1} (data from Evans, 1983). Model lines generated from Eq. [2] and [3].

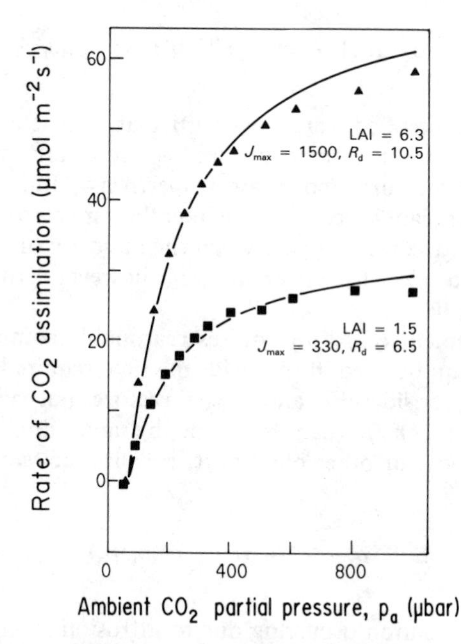

Fig. 1-4. The response of *Triticum aestivum* canopy CO_2 assimilation rate to changes in the ambient partial pressure of CO_2, 80% light interception, Γ_* = 30 μbar, p_c/p_a = 0.67 (data from Evans, 1983). Model lines generated from Eq. [2] and Eq. [3] with irradiance of 1000 μmol quanta m^{-2} s^{-1}.

fall below the curves at high pCO_2, where single-leaf data have also been found to deviate from Eq. [3] (Woo & Wong, 1983). Since these conditions are irrelevant to the majority of leaf canopies, at least at the present ambient pCO_2, the deviations are of little consequence.

While it is relatively easy to define the potential rate of canopy photosynthesis in this way, environmental stresses will reduce the actual rate of photosynthesis, and these effects are more difficult to model. Also, in the field, the light environment is composed of both direct and diffuse irradiance. As the relative proportions of these change, irradiance response curves of the crop are likely to be affected (Kumura, 1968). To link the rate of CO_2 assimilation of leaves to growth also requires knowledge of the considerable respiratory losses and how C is partitioned between leaves and other organs.

TRANSPIRATION-USE EFFICIENCY AND CARBON-ISOTOPE DISCRIMINATION

At the level of single leaves, the transpiration-use efficiency, W, depends on the leaf-to-air water vapor-pressure difference and the ratio of intercellular pCO_2 to ambient pCO_2 (p_i/p_a).

$$W = (1 - \phi_c)\, p_a(1 - p_i/p_a)/[1.6(1 + \phi_w)(e_i - e_a)] \qquad [7]$$

where ϕ_w is the uncontrolled water loss (cuticular, soil, etc.) as a proportion of daytime stomatal transpiration, and e_i and e_a are the water vapor pressures in the leaf and surrounding air, respectively (Hubick & Farquhar, 1989). Obviously, the plant breeder cannot alter the vapor-pressure difference apart from changing the time of year when the majority of growth occurs. If it were possible to select for different p_i/p_a, however, progress could be made in improving W.

Direct measurement of p_i/p_a requires reasonably complicated equipment. Although it can be used in the field, the time required to resolve the small differences is considerable and it is, therefore, not very practicable. Indirect screening for p_i/p_a can be made by analyzing the C-isotope composition of the leaf or other plant part, because the two are related as follows:

$$\Delta = a - d + (b - a)(p_i/p_a) \qquad [8]$$

where a is the fractionation occurring due to diffusion in air (4.4‰), d is a term to account for discrimination due to dissolution of CO_2, liquid-phase diffusion, and respiration (1–3‰), and b is the net fractionation caused by Rubisco and PEP carboxylation (~27‰) (see Farquhar et al., 1982; Farquhar

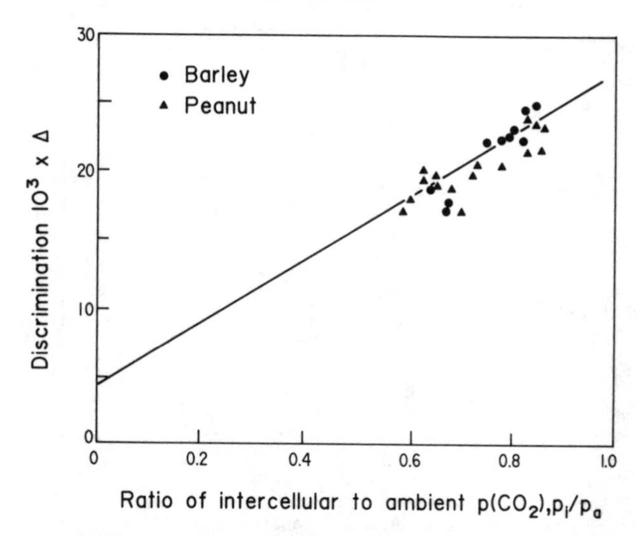

Fig. 1–5. The relationship between concurrent measurement of C isotope discrimination and the ratio of intercellular to ambient partial pressure of CO_2. The theoretical line (Eq. [8]) is shown with values of $a = 4.4$ ‰, $b = 27$ ‰, and $d = 0$ (redrawn from Farquhar et al., 1988).

& Richards, 1984; Evans et al., 1986; Hubick et al., 1986). The validity of Eq. [8] has been shown by experiments in which the change in the isotopic composition of air was measured after it passed over a photosynthesizing leaf (Fig. 1–5; Evans et al., 1986; Farquhar et al., 1988). By substituting Eq. [8] into Eq. [7], the relationship between W and Δ is obtained:

$$W = (1 - \phi_c)\, p_a(b - d - \Delta)/[1.6(1 + \phi_w)(b - a)(e_i - e_a)] \qquad [9]$$

The validity of Eq. [9] has been demonstrated at the plant level for several species. Figure 1–6 shows data for barley (*Hordeum vulgare* L.) (Hubick & Farquhar, 1989), and similar correlations have been obtained with peanut (*Arachis* spp.), wheat (*T. aestivum*), sunflower (*Helianthus annuus* L.), and tomato (*Lycopersicon* spp.) (Martin & Thorstenson, 1988). Initial screening has revealed genetic variation that is heritable, so measuring Δ holds promise as a valuable tool for plant selection (Farquhar et al., 1988). While the results are consistent with Eq. [9] at the leaf and pot level, problems can be anticipated when scaling up to large crop areas if the source of variation is stomatal (Farquhar et al., 1988). This arises because the canopy upwind alters the water vapor content and temperature of the air, thereby affecting W downwind. While W may be higher at the leading edge of the field, the gain may be more marginal downwind. These problems are currently under investigation. It appears that this may be a problem for barley and wheat, but not for sunflower (Virgona et al., 1990) or peanut (Wright et al., 1988; Hubick et al., 1988), where the majority of variation is in photosynthetic capacity.

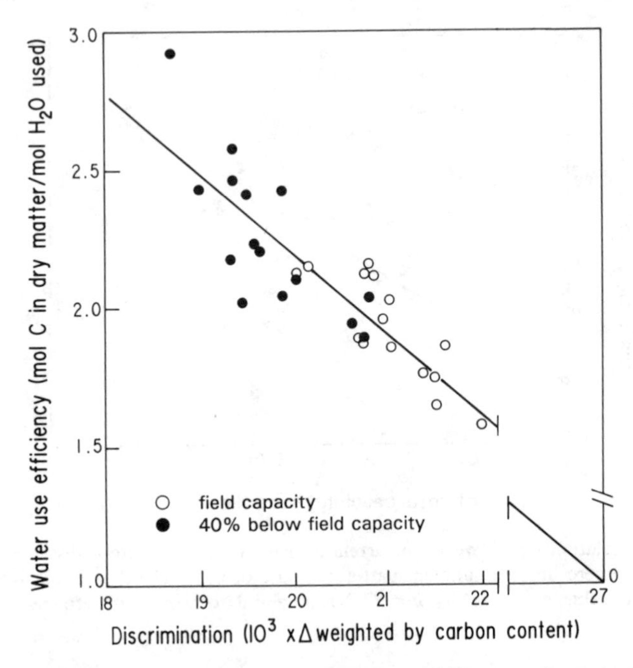

Fig. 1-6. The relationship between water-use efficiency and C isotope discrimination of barley plants grown in pots. Five different genotypes were compared (r^2 = 0.87, redrawn from Hubick & Farquhar, 1989).

CONCLUSION

The model of Farquhar et al. (1980) describes the photosynthetic properties of leaves in terms of the underlying biochemistry and provides a good description of many characteristics of steady-state photosynthesis. The equations given here represent the main features without the derivations or extensions that accommodate, for example, Rubisco activation and RuBP pool sizes. The irradiance-response curve requires empirical fitting to determine the curvature factor, which, at present, cannot be predicted from the biochemistry. However, the exact shape of the irradiance-response curves of leaves becomes less important when incorporated into a big-leaf model, as fine detail becomes obscured.

The equations enable the rate of CO_2 assimilation to be calculated as a function of irradiance and pCO_2. They can incorporate a temperature dependency (see Ch. 2 in this book, Harley & Tenhunen). Despite making gross simplifications in treating a canopy as a big leaf, responses to both irradiance and CO_2 could be predicted for wheat crops grown in a controlled environment. This is presumably because irradiance is the major determinant of canopy photosynthesis. The Rubisco kinetics, then, simply describe the balance between CO_2 assimilation and photorespiration. Obviously the simplifications result in less sensitivity for the big-leaf model than more sophisticated descriptions of the canopy can provide (see Ch. 4, 5, & 6 in this book, by Gutschick, Norman, and Sinclair, respectively).

REFERENCES

Anderson, J.M. 1986. Photoregulation of the composition, function and structure of thylakoid membranes. Annu. Rev. Plant Physiol. 37:93–136.

Björkman, O., and B. Demmig. 1987. Photon yield of O_2 evolution and chlorophyll fluorescence characteristics at 77K among vascular plants of diverse origins. Planta 170:489–504.

Charles-Edwards, D.A., H. Stutzel, R. Ferraris, and D.F. Beech. 1987. An analysis of spatial variation in the nitrogen content of leaves from different horizons within a canopy. Ann. Bot. (London) 60:421–426.

Condon, A.G., R.A. Richards, and G.D. Farquhar. 1987. Carbon isotope discrimination is positively correlated with grain yield and dry matter production in field-grown wheat. Crop Sci. 27:996–1001.

DeJong, T.M., and J.F. Doyle. 1985. Seasonal relationships between leaf nitrogen content (photosynthetic capacity) and leaf canopy light exposure in peach (*Prunus persica*). Plant Cell Environ. 8:701–706.

Delieu, T., and D.A. Walker. 1981. Polarographic measurement of photosynthetic oxygen evolution by leaf discs. New Phytol. 89:165–178.

Evans, J.R. 1983. Photosynthesis and nitrogen partitioning in leaves of *Triticum aestivum* and related species. Ph.D. diss., Australian National Univ., Canberra.

Evans, J.R. 1985. A comparison of the photosynthetic properties of flag leaves of *Triticum aestivum* and *T. monococcum*. p. 111–125. *In* B. Jeffcoat et al. (ed.) British Plant Growth Regulator Group, Bristol.

Evans, J.R. 1986. The relationship between carbon-dioxide-limited photosynthetic rate and ribulose-1,5-bisphosphate carboxylase content in two nuclear-cytoplasm substitution lines of wheat, and the coordination of ribulose-1,5-bisphosphate-carboxylation and electron-transport capacities. Planta 167:351–358.

Evans, J.R. 1987a. The dependence of quantum yield on wavelength and growth irradiance. Aust. J. Plant Physiol. 14:69–79.

Evans, J.R. 1987b. The relationship between electron transport components and photosynthetic capacity in pea leaves grown at different irradiances. Aust. J. Plant Physiol. 14:157–170.

Evans, J.R. 1988. Acclimation by the thylakoid membranes to growth irradiance and the partitioning of nitrogen between soluble and thylakoid proteins. p. 93–106. *In* J.R. Evans et al. (ed.) Ecology of photosynthesis in sun and shade. CSIRO, Melbourne, Australia.

Evans, J.R., T.D. Sharkey, J.A. Berry, and G.D. Farquhar. 1986. Carbon isotope discrimination measured concurrently with gas exchange to investigate CO_2 diffusion in leaves of higher plants. Aust. J. Plant Physiol. 13:281–292.

Evans, J.R., and I. Terashima. 1987. Effects of nitrogen nutrition on electron transport components and photosynthesis in spinach. Aust. J. Plant Physiol. 14:59–68.

Evans, J.R., and I. Terashima. 1988. Photosynthetic characteristics of spinach leaves grown with different nitrogen treatments. Plant Cell Physiol. 29:157–165.

Farquhar, G.D. 1979. Model describing the kinetics of ribulose bisphosphate carboxylase oxygenase. Arch. Biochem. Biophys. 193:456–468.

Farquhar, G.D. 1989. Models of integrated photosynthesis of cells and leaves. Philos. Trans. R. Soc. London. 323:357–368.

Farquhar, G.D., K.T. Hubick, A.G. Condon, and R.A. Richards. 1988. Carbon isotope fractionation and plant water-use efficiency. p. 21–40. *In* P.W. Rundel et al. (ed.) Applications of stable isotope ratios to ecological research. Springer-Verlag, Berlin.

Farquhar, G.D., M.H. O'Leary, and J.A. Berry. 1982. On the relationship between carbon isotope discrimination and the intercellular carbon dioxide concentration in leaves. Aust. J. Plant Physiol. 9:121–137.

Farquhar, G.D., and R.A. Richards. 1984. Isotopic composition of plant carbon correlates with water-use efficiency of wheat genotypes. Aust. J. Plant Physiol. 11:539–552.

Farquhar, G.D., and S. von Caemmerer. 1982. Modelling of photosynthetic response to environmental conditions. p. 550–587. *In* O.L. Lange et al. (ed.) Encycl. Plant Physiol. New Ser. Vol. 12B. Springer-Verlag, Berlin.

Farquhar, G.D., S. von Caemmerer, and J.A. Berry. 1980. A biochemical model of photosynthetic CO_2 assimilation in leaves of C_3 species. Planta 149:78–90.

Hirose, T., and M.J.A. Werger. 1987. Maximising daily canopy photosynthesis with respect to the leaf nitrogen allocation pattern in the canopy. Oecologia 72:520–526.

Hubick, K.T., and G.D. Farquhar. 1989. Carbon isotope discrimination and the ratio of carbon gained to water lost in barley cultivars. Plant Cell Environ. 12:795–804.

Hubick, K.T., G.D. Farquhar, and R [illegible] r-use efficiency and carbon isotope discriminati [illegible]. Aust. J. Plant Physiol. 13:803–816.

Hubick, K.T., R. Shorter, and G.D [illegible] ype × environment interactions of carbon isot [illegible] ency in peanut, *Arachis hypogaea* L. Aust. J. I [illegible]

Inada, K. 1976. Action spectra for pho [illegible] iol. 17:355–365.

King, R.W., and L.T. Evans. 1967. F [illegible] wheat, lucerne and subterranean clover plants. [illegible]

Kobza, J., and J.R. Seemann. 1988 [illegible] ion of ribulose 1, 5-bisphosphate carboxylase act [illegible] oc. Natl. Acad. Sci. (USA) 85:3815–3819.

Koh, S., and A. Kumura. 1975. Studies on matter production in wheat plants II. Carbon dioxide balance and efficiency of solar energy utilization in wheat stand. Proc. Crop Sci. Soc. Jpn. 44:335–342.

Kumura, A. 1968. Studies on dry matter production of soybean plant. 3. Photosynthetic rate of soybean plant population as affected by proportion of diffuse light. Proc. Crop Sci. Soc. Jpn. 37:570–582.

Labate, C.A., and R.C. Leegood. 1988. Limitation of photosynthesis by changes in temperature. Planta 173:519–527.

Leverenz, J.W. 1987. Chlorophyll content and the light response curve of shade-adapted conifer needles. Physiol. Plant. 71:20–29.

Leverenz, J.W. 1988. The effects of illumination sequence, CO_2 concentration, temperature and acclimation on the convexity of the photosynthetic light response curve. Physiol. Plant. 74:332–341.

Martin, B., and Y.R. Thorstenson. 1988. Stable carbon isotope composition ($\delta^{13}C$), water use efficiency and biomass productivity of *Lycopersicon esculentum*, *Lycopersicon pennellii* and the F_1 hybrid. Plant Physiol. 88:213–217.

Masle, J., and G.D. Farquhar. 1988. Effects of soil strength on the relation of water-use efficiency and growth to carbon isotope discrimination in wheat seedlings. Plant Physiol. 86:32–38.

McCree, K.J. 1972. The action spectrum, absorptance and quantum yield of photosynthesis in crop plants. Agric. Meteorol. 9:191–216.

McCree, K.J. 1986. Whole plant carbon balance during osmotic adjustment to drought and salinity stress. Aust. J. Plant Physiol. 13:33–44.

Oja, V.M., and A.K. Laisk. 1976. Adaptation of the photosynthesis apparatus to the light profile in the leaf. Sov. Plant Physiol. (Engl. Transl.) 23:381–386.

Sage, R.F., and T.D. Sharkey. 1987. The effects of temperature on the occurrence of O_2 and CO_2 insensitive photosynthesis in field grown plants. Plant Physiol. 84:658–664.

Sharkey, T.D. 1985. Photosynthesis in intact leaves of C_3 plants: Physics, physiology and limitations. Bot. Rev. 51:53–105.

Sharkey, T.D., J.R. Seemann, and J.A. Berry. 1986. Regulation of ribulose 1,5-bisphosphate carboxylase activity in response to changing partial pressure of O_2 and light in *Phaseolus vulgaris*. Plant Physiol. 81:788–791.

Terashima, I. 1986. Dorsiventrality in photosynthetic light response curves of a leaf. J. Exp. Bot. 37:399–405.

Terashima, I., and J.R. Evans. 1988. Effects of light and nitrogen nutrition on the organization of the photosynthetic apparatus in spinach. Plant Cell Physiol. 29:143–155.

Terashima, I., and Y. Inoue. 1985. Vertical gradients in photosynthetic properties of spinach chloroplasts dependent on intraleaf light environment. Plant Cell Physiol. 26:781–785.

Terashima, I., and T. Saeki. 1983. Light environment within a leaf I. Optical properties of paradermal sections of *Camellia* leaves with special reference to differences in the optical properties of palisade and spongy tissues. Plant Cell Physiol. 24:1493–1501.

Terashima, I., and T. Saeki. 1985. A new model for leaf photosynthesis incorporating the gradients of light environment and of leaf photosynthetic properties of chloroplasts within a leaf. Ann. Bot. (London) 56:489–499.

Virgona, J.M., K.T. Hubick, H.M. Rawson, G.D. Farquhar, and R.W. Downes. 1990. Genotypic variation in transpiration efficiency and carbon allocation during early growth in sunflower. Aust. J. Plant Physiol. 17:207–214.

von Caemmerer, S., and D.L. Edmonson. 1986. Relationship between steady-state gas exchange in vivo ribulose bisphosphate carboxylase activity and some carbon reduction cycle intermediates in *Raphanus sativus*. Aust. J. Plant Physiol. 13:669–688.

von Caemmerer, S., and G.D. Farquhar. 1981. Some relationships between the biochemistry of photosynthesis and the gas exchange of leaves. Planta 153:376–387.

Walters, M.B., and C.B. Field. 1987. Photosynthetic light acclimation in two rainforest *Piper* species with different ecological amplitudes. Oecologia 72:449–456.

Woo, K.C., and S.C. Wong. 1983. Inhibition of CO_2 assimilation by supraoptimal CO_2: Effect of light and temperature. Aust. J. Plant Physiol. 10:75–85.

Wright, G.C., K.T. Hubick, and G.D. Farquhar. 1988. Discrimination in carbon isotopes of leaves correlates with water-use efficiency of field grown peanut cultivars. Aust. J. Plant Physiol. 15:815–825.

2 Modeling the Photosynthetic Response of C_3 Leaves to Environmental Factors

P. C. Harley and J. D. Tenhunen
San Diego State University
San Diego, CA

The primary substrate required for plant growth is the C acquired in the process of photosynthesis, some of which is lost as respiration, and the rest of which is partitioned to various organs. In theory, all leaves in a plant canopy could be treated individually, and their rates of CO_2 uptake summed to estimate the pool of newly fixed C available to support plant growth and maintenance. In practice, of course, leaves may be aggregated in one of several ways, and average responses for these groups of leaves are summed for the canopy as a whole. In this process, individual leaf properties assume less importance than the properties of each layer or of the entire canopy. Thus, whereas maximum carboxylation capacity may be an important parameter in determining individual leaf behavior, canopy light-utilization efficiency is far more valuable for predicting canopy gas-exchange behavior. The premise of this chapter, however, is that an understanding of the way in which individual leaves interact with their environment can lead to new insights into plant canopy gas-exchange behavior. The best way to understand the complex interactions between leaves and their environment is by the use of a mechanistically based model of leaf gas exchange. The model discussed below describes the photosynthetic responses of leaves of C_3 plants as affected by incident irradiance, leaf temperature, and the concentrations of CO_2 and O_2.

THE C_3 LEAF PHOTOSYNTHESIS/CONDUCTANCE MODEL

Although CO_2 fixation is a complex process, dependent on light energy absorption and transduction and a series of biochemical steps, a convincing case can be made that the photosynthetic responses of C_3 leaves may be interpreted based solely on the kinetics of a single enzyme, ribulose-1,5-bisphosphate carboxylase-oxygenase (Rubisco) (Woodrow & Berry, 1988). The rate of carboxylation is determined by the amount and kinetic properties of Rubisco, and also depends on (i) the concentrations of competing gaseous substrates, CO_2 and O_2, (ii) the ratio of ribulose-1,5-bisphosphate (RuBP) concentration to enzyme active sites, (iii) the concentration of the inhibitor carboxyarabinitol-1-P, and (iv) the concentrations of various effectors that influence the activation state of Rubisco (Woodrow & Berry, 1988).

The model of C_3 photosynthesis presented here was proposed by Farquhar et al. (1980) and is based on Rubisco kinetics as mediated by Factors (i) and (ii) above. Because the reaction is ordered with RuBP binding first, and because the concentration of RuBP binding sites on Rubisco is nearly two orders of magnitude greater than the Michaelis constant (K_m) for RuBP, Rubisco fails to obey typical kinetics with respect to the substrate, RuBP (Farquhar, 1979). To a good approximation, the rate of carboxylation is either RuBP saturated, in which case it is determined by the kinetic properties of the enzyme and the concentrations of CO_2 and O_2, or RuBP limited and determined by the rate of regeneration of RuBP in the integrated Calvin and photorespiratory cycles (see Farquhar & von Caemmerer [1982] and Sharkey [1985a] for details). The regeneration of RuBP is assumed to be dependent on the rate of adenosine triphosphate (ATP) supply, which is usually limited by the electron-transport capacity. Under certain conditions, however, RuBP regeneration is limited by the availability of inorganic phosphate (P_i) for photophosphorylation, which in the short term is determined by the relative rates of P_i sequestration in the production of triose phosphates and P_i regeneration, as triose phosphates are utilized in starch and sucrose production (Sharkey, 1985a).

Given that one molecule of CO_2 is released in photorespiration for every two oxygenations of RuBP, net photosynthesis (A) may be expressed as

$$A = V_c - 0.5V_o - R_d = V_c\left(1 - 0.5\frac{V_o}{V_c}\right) - R_d \qquad [1]$$

where V_c and V_o are the rates of RuBP carboxylation and oxygenation respectively, and R_d is *day respiration*, the rate of CO_2 evolution from processes other than photorespiration (i.e., mitochondrial respiration from nonautotrophic tissue or mitochondrial respiration in autotrophic cells that continues in the light). Assuming Michaelis-Menten competitive kinetics, V_o/V_c may be expressed in terms of the enzyme specificity factor, τ,

$$V_o/V_c = (1/\tau)\, O/C_i \qquad [2]$$

where O and C_i are the concentrations of O_2 and CO_2 at the site of fixation. The factor τ may be expressed in terms of enzyme kinetic parameters as follows:

$$\tau = (V_{c_{max}} K_o)/(V_{o_{max}} K_c) \qquad [3]$$

where $V_{c_{max}}$ and $V_{o_{max}}$ are maximum rates and K_c and K_o are Michaelis-Menten constants for carboxylation and oxygenation, respectively. In order to remain consistent with the terminology developed by Farquhar, we now introduce Γ_*, the CO_2 compensation point in the absence of mitochandrial respiration, which may also be expressed in terms of kinetic parameters and is inversely related to τ, according to the following relationship:

$$\Gamma_* = \frac{0.5\ V_{o_{max}}\ K_c}{V_{c_{max}}\ K_o} O = \frac{0.5\ O}{\tau}$$

Making the appropriate substitutions, Eq. [1] may be rewritten

$$A = V_c\left(1 - \frac{\Gamma_*}{C_i}\right) - R_d \qquad [5]$$

As discussed above, V_c is limited either by the activity of Rubisco (assuming saturating amounts of RuBP) or by the rate of regeneration of RuBP, mediated either by the electron-transport rate or P_i. Thus, $V_c = \min\{W_c, W_j, W_p\}$ where min{} denotes minimum of, W_c is the Rubisco-limited rate of carboxylation, and W_j and W_p are the RuBP-limited rates of carboxylation when RuBP regeneration is limited by electron transport and P_i, respectively. Substituting,

$$A = \left(1 - \frac{\Gamma_*}{C_i}\right) \min\{W_c, W_j, W_p\} - R_d \qquad [6]$$

The rate W_c is assumed to obey competitive Michaelis-Menten kinetics; thus,

$$W_c = \frac{V_{c_{max}}\ C_i}{C_i + K_c(1 + O/K_o)}. \qquad [7]$$

The rate W_o may be expressed analogously:

$$W_o = \frac{V_{o_{max}}\ O}{O + K_o(1 + C_i/K_c)} = \frac{W_c\ \Gamma_*}{0.5C_i}. \qquad [8]$$

The rates W_c and W_o assume saturating amounts of RuBP, which is often not the case, inasmuch as RuBP concentration is dependent on the balance between RuBP utilization in carboxylation and oxygenation, and RuBP regeneration in the Calvin cycle. Farquhar and von Caemmerer (1982) developed several formulations for W_j, the electron-transport-limited rate of carboxylation, depending on whether $NADPH^+$ or ATP limits Calvin cycle activity, but we adopt a slightly more empirical formulation. Our parameter P_m is the CO_2-saturated rate of photosynthesis at any given irradiance and temperature. Under these RuBP-limited conditions, each RuBP molecule is utilized immediately as it becomes available and P_m thus represents the maximum potential rate of RuBP production. This value is compared with the rate of RuBP utilization under RuBP-saturated conditions ($W_c + W_o$). If $P_m < (W_c + W_o)$, RuBP is limiting, and the ratio $P_m/(W_c + W_o)$ represents the extent of that limitation. Thus, W_c is not attainable, and W_j may be expressed

$$W_j = W_c \frac{P_m}{W_c + W_o} = \frac{P_m}{1 + 2\Gamma_*/C_i}. \quad [9]$$

This expression for W_j is equivalent to that used by Farquhar and von Caemmerer (1982) if their parameter J, the potential rate of electron transport, is equal to $4P_m$; i.e., if four electrons are required for the regeneration of a single RuBP molecule in the Calvin cycle, which is the case when $NADPH^+$ is limiting. Regardless of whether ATP or $NADPH^+$ limits Calvin cycle activity, Eq. [9] remains valid as long as P_i is not limiting. Note that when RuBP regeneration is limited by electron transport, CO_2 and O_2 continue to compete for active sites and assimilation remains sensitive to CO_2.

In contrast, when RuBP regeneration is limited by the availability of P_i, assimilation becomes insensitive to both CO_2 and O_2 (Sharkey, 1985b); in fact, this is the only way in which P_i limitations can be inferred from gas-exchange measurements. Under these conditions, photosynthesis is controlled by the rate of P_i release as triose phosphate is utilized. Given that stromal P_i will decline unless triose phosphate is utilized at one-third the rate of carboxylation (Walker & Herold, 1977), Sharkey (1985a) proposed that, under P_i-limited conditions, A was equal to 3 TPU $- R_d$, where TPU is the rate of triose phosphate utilization. In the context of this model, therefore,

$$W_p = 3\ \mathrm{TPU} / \left(1 - \frac{\Gamma_*}{C_i}\right). \quad [10]$$

The relationship between the RuBP-limited and RuBP-saturated portions of the A vs. C_i response, as defined by Eq. [6] through [10], is shown in Fig. 2–1, which describes model responses at both high and low irradiance. In both cases, RuBP is saturating at low C_i and assimilation is limited by Rubisco ($V_c = W_c$). As C_i increases, a point (labeled *Transition*) is reached at which RuBP becomes limiting. At relatively low irradiance (bottom panel) electron transport (ET) limits RuBP regeneration ($V_c = W_j$) and observed assimilation (solid line) remains slightly sensitive to C_i. At high irradiance, potential electron-transport rates exceed the capacity of starch and sucrose synthesis to sustain chloroplast levels of P_i, which then limits carboxylation ($V_c = W_p$). (It has been noted that the transition from one limitation to another, which is observed experimentally, is rarely as abrupt

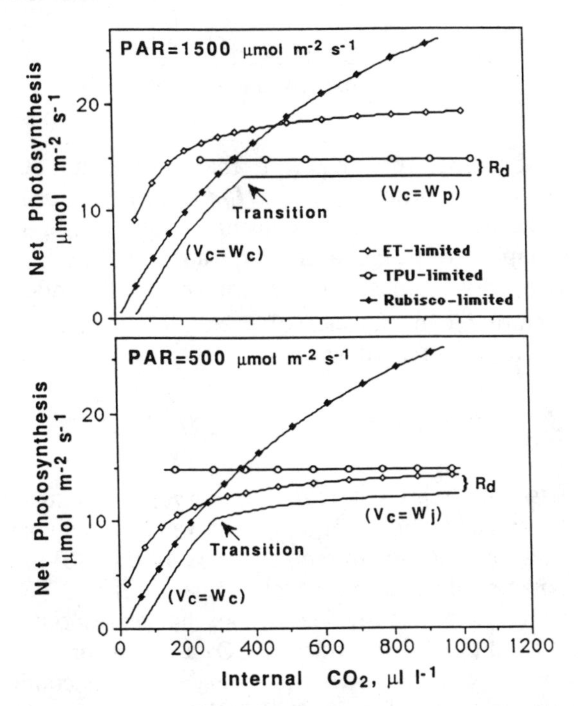

Fig. 2-1. Schematic description of model behavior simulated at high light (above) and low light (below). Net photosynthesis (no symbols) is the minimum of three functions that are limited by Rubisco, electron transport (ET), and triose phosphate utilization (TPU). R_d is respiration continuing in the light.

as that in Fig. 2-1, presumably due to the nature of enzyme kinetics and nonhomogeneity of physiological properties within the leaf [Kirschbaum & Farquhar, 1984]. Kirschbaum and Farquhar include an empirical smoothing factor to make this transition more gradual.)

At a given irradiance and temperature, therefore, net photosynthesis may be described using seven parameters, the three kinetic parameters $V_{c_{max}}$, K_c, and K_o, the specificity factor τ, P_m, TPU, and R_d.

We describe a light dependency only for P_m, using the following equation (Smith, 1937; Tenhunen et al., 1976):

$$P_m = \frac{\alpha I}{\left(1 + \dfrac{\alpha^2 I^2}{P_{ml}^2}\right)^{1/2}} \qquad [11]$$

where I is incident photosynthetic photon flux density, α is the initial slope of the curve relating CO_2-saturated photosynthesis to irradiance, analogous to quantum use efficiency (on an incident-light basis), and P_{ml} is the CO_2- and light-saturated rate of photosynthesis. The temperature dependency of P_{ml} is described as follows (Johnson et al., 1942; Tenhunen et al., 1976):

$$P_{ml} = \frac{\exp(c - \Delta H_a/RT_K)}{1 + \exp[(\Delta S T_K - \Delta H_d)/RT_K]} \quad [12]$$

where T_K is leaf temperature (K), R is the gas constant (0.0083 J K^{-1} mol^{-1}), ΔH_a is the activation energy, ΔH_d the energy of deactivation, ΔS is an entropy term and c is a scaling constant. This function has been used to describe uncoupled whole-chain electron transport (Farquhar et al., 1980) and will be used below to describe the temperature dependency of $V_{c_{max}}$.

The temperature dependencies of τ, K_c, K_o, and R_d may each be described by an exponential function

$$\text{Parameter} = \exp(c - \Delta H_a/RT_K) \quad [13]$$

where *Parameter* may represent τ, K_c, K_o, or R_d, and c and ΔH_a are the scaling constant and activation energy, respectively, for each parameter. The value of TPU also increases with temperature and Eq. [13] may apply, but no data yet exist on which to base such a relationship.

To be useful in predicting leaf gas-exchange responses to varying environmental conditions, the model of CO_2 assimilation presented above must be integrated with a model describing stomatal conductance (g_s). Although the physiology of stomata has been extensively examined, a mechanistic understanding of how stomata respond to light, temperature, humidity, and CO_2 remains elusive. The often-noted correlation between stomatal conductance and net CO_2 assimilation (Wong et al., 1979) has led to the development of empirical models in which assimilation is one parameter used to predict conductance. Ball et al. (1987) developed the following empirical model to describe stomatal conductance:

$$g_s = k\,A\,h_s/C_s \quad [14]$$

where h_s and C_s are relative humidity (as a decimal fraction) and CO_2 concentration, respectively, at the leaf surface, and k is a constant representing stomatal sensitivity to these factors.

Stomatal conductance is thus dependent on A, while A is dependent on g_s through the latter's effect on C_i, according to the equation

$$C_i = C_a - 1.6\,A/g_s \quad [15]$$

where C_a is the concentration of CO_2 external to the leaf, and the factor 1.6 accounts for the different diffusivities of water vapor and CO_2 in air. Given these interdependencies, the model must solve for C_i iteratively. Thus, A and g_s are integrated by iterating for that value of C_i that is compatible with both the photosynthetic rate (Eq. [6]) and conductance (Eq. [14]) predicted by the models, where A and g_s are related according to Eq. [15].

Table 2-1. List of parameters used in the model, the equations in which they are used, and the units of each. For those parameters considered to be constant, values are also given.

Parameter	Equation	Value	Units
$c(P_{ml})$		--	--
$\Delta H_a(P_{ml})$	[12]	--	$kJ\ mol^{-1}$
$\Delta H_d(P_{ml})$		--	$kJ\ mol^{-1}$
$\Delta S(P_{ml})$		--	$kJ\ K^{-1}\ mol^{-1}$
α	[11]	(0.06)	$\frac{mol\ CO_2}{mol\ photon}$
$c(R_d)$	[13]	--	--
$\Delta H_a(R_d)$		--	$kJ\ mol^{-1}$
$c(K_c)$	[13]	31.95	--
$\Delta H_a(K_c)$		65.0	$kJ\ mol^{-1}$
$c(K_o)$	[13]	19.61	--
$\Delta H_a(K_o)$		36.0	$kJ\ mol^{-1}$
$c(\tau)$	[13]	−3.949	--
$\Delta H_a(\tau)$		−28.99	$kJ\ mol^{-1}$
$c(V_{c_{max}})$		--	--
$\Delta H_a(V_{c_{max}})$	[12]	--	$kJ\ mol^{-1}$
$\Delta H_d(V_{c_{max}})$		--	$kJ\ mol^{-1}$
$\Delta S(V_{c_{max}})$		--	$kJ\ K^{-1}\ mol^{-1}$
TPU	[10]	--	$\mu mol\ m^{-2}\ s^{-1}$
k	[14]	--	--

MODEL PARAMETERIZATION

For a complete parameterization of the model, the temperature dependencies of P_{ml} and $V_{c_{max}}$ (Eq. [12]), of K_c, K_o, R_d, and τ (Eq. [13]), and of α and TPU need to be determined. The complete set of model parameters requiring estimation, and their respective units, is given in Table 2-1.

Jordan and Ogren (1984) and Brooks and Farquhar (1985) demonstrated that τ decreased (and Γ_* increased) with temperature similarly for all C_3 species examined, whether determined from enzyme assays in vitro or from gas-exchange techniques (see also Woodrow & Berry, 1988). This dependency may be described by Eq. [13], as shown in Fig. 2-2, using the parameter values shown.

To describe the temperature dependencies of K_c and K_o, we have used the data of Badger and Collatz (1977); their values were also fit to Eq. [13], using the energies of activation shown in Fig. 2-2. (Although Badger and Collatz observed a discontinuity in the temperature dependency of K_c at $\approx$ 15°C, we've fit their data to a single exponential function.) Small but significant differences in K_c have been reported from different C_3 plants (Keys, 1986; Evans & Seeman, 1984). Values for K_c (in $\mu L\ L^{-1}$) and K_o (in $mL\ L^{-1}$) at 25°C used in recent modeling efforts include the following: $K_c = 300$, $K_o = 161$ (Farquhar & Wong, 1984); $K_c = 310$, $K_o = 155$

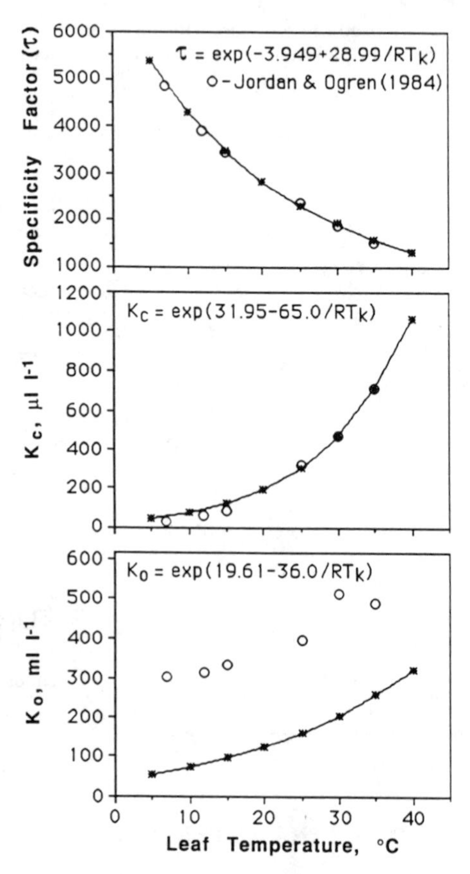

Fig. 2–2. Descriptions of the temperature dependencies of the specificity factor (τ) and the Michaelis constants for carboxylation (K_c) and oxygenation (K_o).

(Kirschbaum & Farquhar, 1984); $K_c = 300$, $K_o = 212$ (von Caemmerer & Farquhar, 1984); and $K_c = 324$, $K_o = 395$ (Jordan & Ogren, 1984; Sage & Sharkey, 1987). This final value of 395 mL L^{-1} for K_o (25°C) is the inhibitory constant for O_2 with respect to CO_2 (designated K_o^i) that Woodrow and Berry (1988) suggested is more accurately assayed than is the K_o of oxygenase activity per se. The appropriate value of K_o needs to be resolved, since it determines the extent to which photorespiration reduces net assimilation. For the present, we have chosen values for $c(K_c)$ and $c(K_o)$ in Eq. [13] that yield values at 25 °C of 305 μL L^{-1} for K_c and 161 mL L^{-1} for K_o (Fig. 2–2).

These temperature dependencies result in oxygenation being increasingly favored over carboxylation with increasing temperature. It has been argued (Ku & Edwards, 1977; Hall & Keys, 1983) that this effect is due, at least in part, to the fact that the ratio of O_2 to CO_2 solubility increases with temperature, and that CO_2 and O_2 are best expressed as molar concentrations in solution. In choosing to express CO_2 and O_2 as concentrations (Badger & Collatz, 1977; Farquhar & von Caemmerer, 1982), we do not deny that these solubility effects may be important. Jordan and Ogren (1984) carefully controlled the gas concentrations in equilibrium with the aqueous phase and concluded that approximately one-third of the observed temperature dependency resulted from gas solubilities. Though the question remains unresolved, solubility effects are incorporated into the temperature dependencies in Fig. 2–2, since K_c and K_o are expressed in terms of concentrations in equilibrium with the dissolved concentrations, and it is a simple matter to convert between the two forms of expression using gas-solubility tables (Hodgman et al., 1958).

The parameter α can only be determined from the initial slope of the light-response curve measured at saturating CO_2. Such measurements are not made routinely, however. Fortunately, α is analogous to quantum yield (although on an incident- rather than absorbed-light basis), which has been shown to be constant among C_3 species when determined under CO_2-saturated conditions (≈ 0.073 mol CO_2 mol^{-1} photon; Ehleringer & Björkman, 1977). Thus, α is expected to be relatively constant (≈ 0.06 mol CO_2 mol^{-1} photon, assuming leaf absorptance $\approx 85\%$), varying only with leaf absorptance characteristics. Although nearly constant under benign conditions, quantum yield is sensitive to various stresses, including high and low temperature (Berry & Bjorkman, 1980), water stress (Mohanty & Boyer, 1976), and photoinhibition (Powles, 1984). In cases where such factors may be important, α should be measured directly.

If we accept the temperature dependencies for K_c, K_o, and τ above, it is possible to estimate values for R_d and $V_{c_{max}}$ at a given temperature from an A vs. C_i response curve. The value of R_d may be determined from the predicted CO_2 release when the initial slope of the response is extrapolated to $C_i = \Gamma_*$, although this requires very precise measurements. Although the initial slope of an A vs. C_i curve is often described as linear, the slope decreases continuously and a linear extrapolation may underestimate R_d unless the lowest point has a C_i value close to Γ_*.

Having estimated R_d, a nonlinear least-squares fit to Eq. [6], which iterates for $V_{c_{max}}$, may be applied using three or more data points on the initial, linear portion of the A vs. C_i response, where $V_c = W_c$ and Eq. [7] applies. If P_i limitations are not apparent (i.e., the plateau portion of the curve is CO_2 sensitive), P_{ml} at a given temperature may be estimated in an analogous fashion, using data from the RuBP-limited portion of the A vs. C_i response, where $V_c = W_j$ and Eq. [9] applies. If an A vs. C_i curve exhibits an abrupt transition and a flat, CO_2-insensitive saturation plateau, P_i is assumed to be limiting, and TPU set equal to $(A + R_d)/3$.

Kirschbaum and Farquhar (1984) established the initial slope of the A vs. C_i response in snowgum (*Eucalyptus pauciflora* Sieb. ex Spreng) for two points measured at $C_i < 100$ μL L^{-1}, from which they estimated both R_d and $V_{c_{max}}$; a single point measured at $C_i \approx 350$ μL L^{-1} was then used to estimate RuBP-regeneration capacity. Although they were successful using this technique, a more complete A vs. C_i curve naturally increases confidence in the parameter estimates.

Although it is impossible to predict a priori where, on the A vs. C_i response, the transition from Rubisco to RuBP limitation occurs, the transitional region is usually apparent in measured data. Farquhar has argued, based on efficient allocation of N to Rubisco and the machinery for RuBP regeneration (Calvin cycle and electron-transport components) that the transition ought to occur near the C_i at which C_3 plants operate (200–250 $\mu L\ L^{-1}$), when determined at the average irradiance experienced by the plant during growth. Supporting evidence comes from von Caemmerer and Farquhar (1984), who found that the zone of transition in common bean (*Phaseolus vulgaris* L.), grown under a variety of conditions, averaged about 225 $\mu L\ L^{-1}$. Harley et al. (1985) reported that the point of transition in soybean, determined at the growth light intensity, was approximately 280 $\mu L\ L^{-1}$. (Note that the measured soybean responses shown in Fig. 2–3 were obtained at saturating irradiance [>2200 $\mu mol\ m^{-2}\ s^{-1}$] rather than the growth irradiance of 800 $\mu mol\ {}^{-2}\ s^{-1}$.)

MODEL PARAMETERIZATION USING A SOYBEAN DATA SET

In this section, the model is parameterized for soybean (*Glycine max* [L.] Merr.), using a portion of a previously published data set (Harley et al., 1985). Figure 2–3 depicts a family of A vs. C_i response curves measured

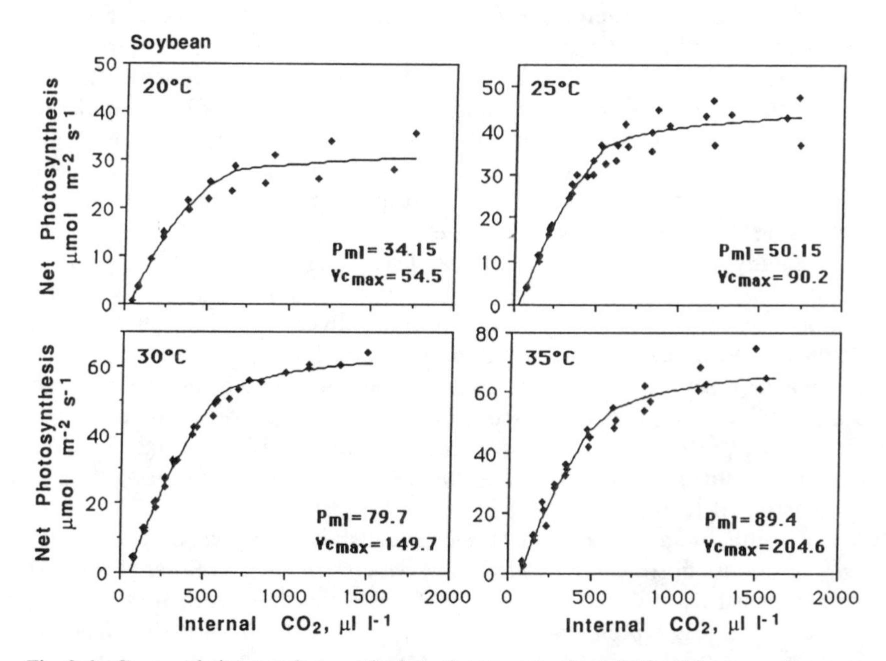

Fig. 2–3. Curves relating net photosynthesis to the concentration of CO_2 (C_i) measured on soybean at four leaf temperatures. Lines are model fits to the data, using the values of P_{ml} and $V_{c_{max}}$ shown.

on soybean at $\approx$2300 μmol photon m^{-2} s^{-1} and over a range of leaf temperatures. Limitations resulting from P_i were apparent only at extremely high C_i values, if at all, and have been ignored. Estimates of $V_{c_{max}}$ and P_{ml} were obtained by least-squares fits to Eq. [6] in conjunction with Eq. [7] and [9] as described above, resulting in the model fits shown (In order to obtain these fits, it was necessary to assume that R_d was zero.)

Data relating P_{ml} and $V_{c_{max}}$ to leaf temperature, based on the fits described in Fig. 2–3, are shown in Fig. 2–4. Solid-line fits for both P_{ml} and $V_{c_{max}}$ were obtained by nonlinear least-squares fit to Eq. [12], resulting in the parameter values shown.

Data relating α to leaf temperature (Harley et al., 1985) indicate that α was constant above 20 °C and equal to $\approx$ 0.06 mol CO_2 mol^{-1} photon.

Finally, a word must be said about estimating k, the slope of the relationship between A and g_s (Eq. [14]), which can only be determined empirically. If g_s has been measured under a variety of temperatures and irradiances, then it may be plotted vs. $A\ h_s/C_s$ and the slope (k) determined (Ball et al., 1987). Such a plot would also allow determination of the validity of the relationship proposed in Eq. [14]. Lacking such a data set, the alternative is to determine the value of k that yields appropriate values of C_i when C_a is 340 μL L^{-1} and light and temperature are near optimal; for most C_3 species, C_i ranges between 200 and 240 μL L^{-1} under such conditions.

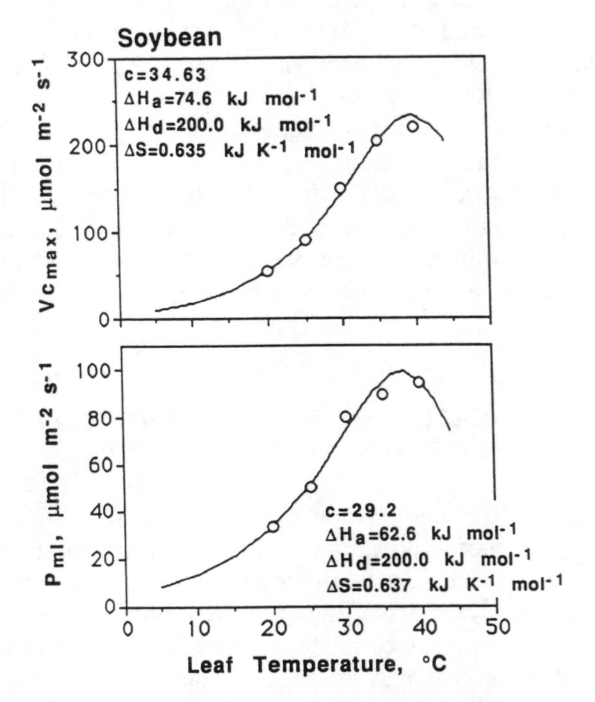

Fig. 2–4. Fits of P_{ml} and $V_{c_{max}}$, obtained in Fig. 2–3, to leaf temperature, using Eq. [12]. Best-fit parameter values are shown (ΔH_d was held constant at 200.0 kJ mol^{-1}).

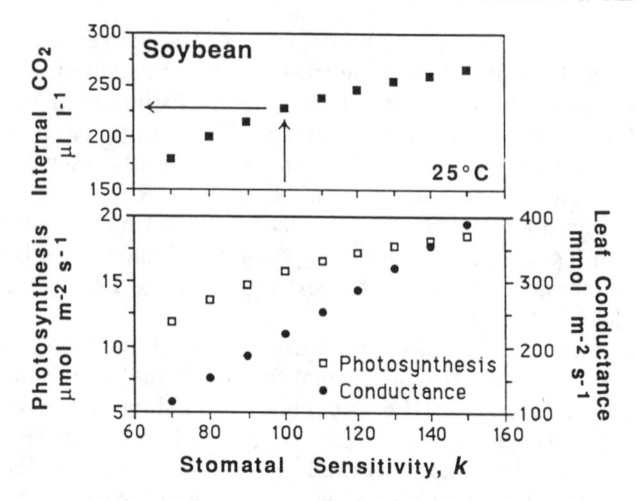

Fig. 2–5. Model simulation of the effects of changing the stomatal sensitivity (k) on leaf conductance, net photosynthesis, and internal concentration of CO_2 (C_i), using soybean model parameters. Simulation is used to determine the value of k that yields reasonable values of C_i (≈ 230 μL L^{-1}).

Figure 2–5 shows the effect of varying k in our parameterized soybean model. As k increases, g_s increases almost linearly, leading to nonlinear increases in both C_i and A. At 25 °C and 1000 μmol photon m^{-2} s^{-1}, C_i for soybean is expected to fall between 220 and 240 μL L^{-1}, which results when k is between 100 and 110.

Although this data set was well suited for model parameterization, and model fits to the parameterization data (Fig. 2-3 and 2-4) were quite good, no independent data were available for these plants against which to test the model. We next parameterize the model for a second species, and attempt to simulate an independently obtained data set describing the diurnal course of net photosynthesis and leaf conductance, measured under field conditions.

APPLICATION OF THE MODEL TO FIELD PHOTOSYNTHETIC RESPONSES

Harley et al. (1986) applied an earlier version of this photosynthesis model to diurnal photosynthetic responses measured on strawberry tree (*Arbutus unedo* L.) in a natural Mediterranean scrub community in Portugal. These data are reanalyzed here using the current photosynthesis model and incorporating the conductance model. Temperature-response curves were obtained for strawberry tree at several CO_2 concentrations, from which A vs. C_i curves at three leaf temperatures were generated (Harley et al., 1986). Figure 2–6 depicts these curves, with model fits obtained using the procedures

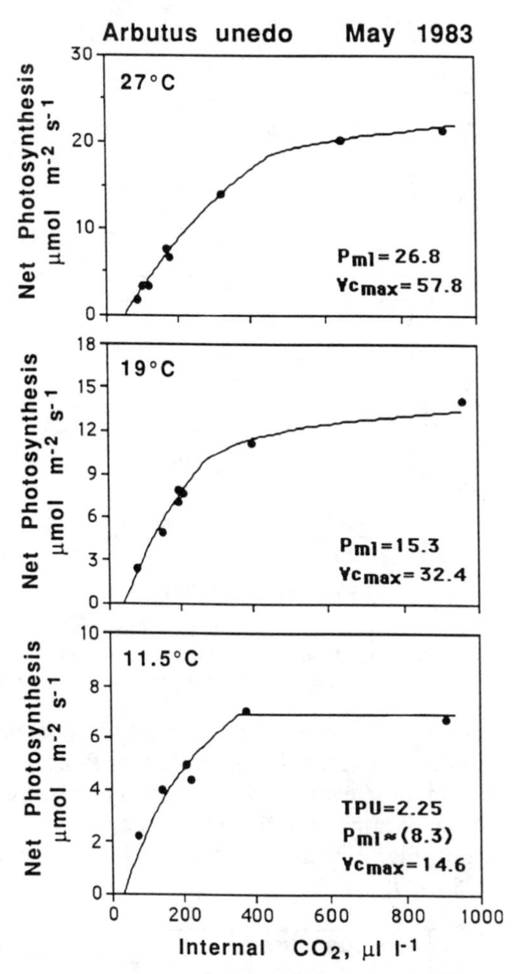

Fig. 2–6. Curves relating net photosynthesis to the concentration of CO_2 (C_i) measured on strawberry tree at three leaf temperatures. Lines are model fits to the data, using the values of P_{ml} and $V_{c_{max}}$ shown.

outlined above. In this instance, however, photosynthesis exhibited apparent CO_2 insensitivity above $\approx 350\ \mu L\ L^{-1}$ at 11.5°C. Triose phosphate utilization was assigned a value of $2.25[(A + R_d)/3]$ on that basis. (Since P_i is assumed to be limiting, we cannot obtain an estimate of P_{ml} at 11.5°C from these data. For the purposes of fitting P_{ml} to a temperature function, however, it was assigned a value of 8.3. This inability to determine a value of P_{ml} when P_i is limiting is a shortcoming of this parameterization protocol, although it could be resolved with difficulty by reducing light until the electron-transport rate became limiting and P_{ml} could be estimated.) The three values of $V_{c_{max}}$ and P_{ml} were then fit to Eq. [12] using nonlinear least-squares methodology (holding $\Delta H_d = 200.0$ kJ mol^{-1}) as shown in Fig. 2–7. (Obviously, since data are lacking at and above the temperature

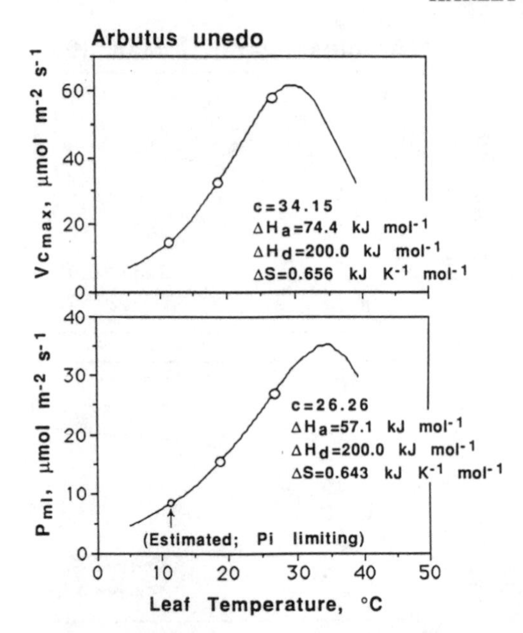

Fig. 2-7. Fits of P_{ml} and $V_{c_{max}}$, obtained in Fig. 2-6, to leaf temperature, using Eq. [12]. Best-fit parameter values are shown (ΔH_d was held constant at 200.0 kJ mol^{-1}). The open circle represents an estimation of P_{ml} under conditions of limiting inorganic P (P_i).

optimum for both $V_{c_{max}}$ and P_{ml}, the model cannot be applied with confidence above ≈27 °C.) Since, in this instance, P_i became limiting only at C_i values well above those experienced at ambient CO_2, we need no longer be concerned with the parameter TPU. The value of k in Eq. [14] was estimated as for soybean (Fig. 2–8) and a value of 100 was used, which yielded a C_i value of ≈225 μL L^{-1}.

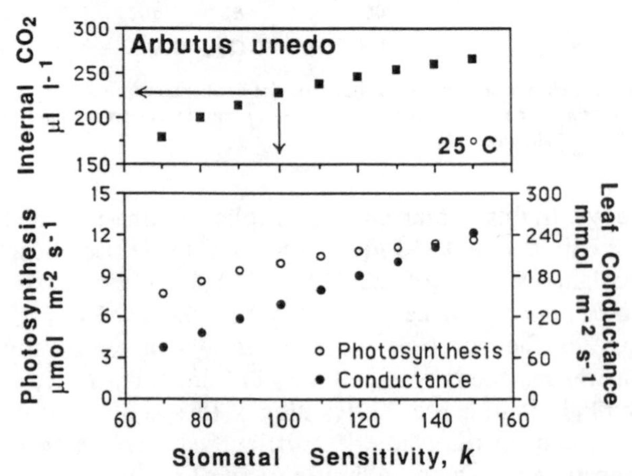

Fig. 2–8. Model simulation of the effects of changing the stomatal sensitivity (k) on leaf conductance, net photosynthesis, and internal concentration of CO_2 (C_i), using strawberry tree model parameters. Simulation is used to determine the value of k that yields reasonable values of C_i (≈225 μL L^{-1}).

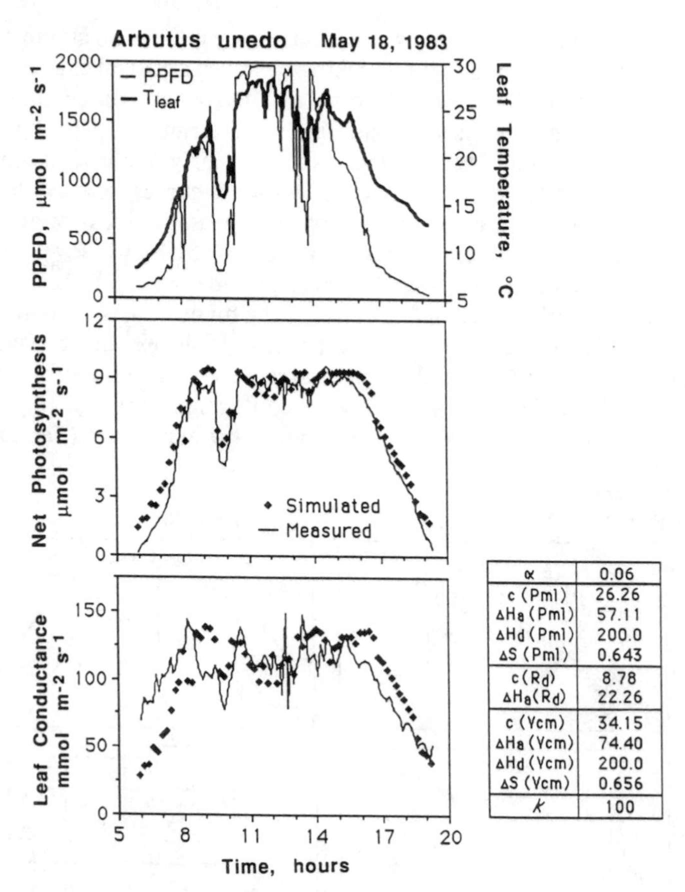

α	0.06
c (Pml)	26.26
ΔHa (Pml)	57.11
ΔHd (Pml)	200.0
ΔS (Pml)	0.643
c(Rd)	8.78
ΔHa(Rd)	22.26
c (Vcm)	34.15
ΔHa (Vcm)	74.40
ΔHd (Vcm)	200.0
ΔS (Vcm)	0.656
k	100

Fig. 2-9. Diurnal responses of net photosynthesis and leaf conductances of strawberry tree, measured in the field in Portugal. Also shown are diurnal courses of light (photosynthetic photon flux density [PPFD]) and leaf temperature. Symbols represent model simulations of the data, using parameter set shown.

Using the parameter values obtained above, the model was used to simulate an independent data set. The diurnal responses of net assimilation and leaf conductance for a strawberry tree shoot were measured under field conditions (Harley et al., 1986) and the environmental inputs used to simulate the data (Fig. 2-9). Despite a tendency to overestimate photosynthesis when irradiance was low early and late in the day, the overall fit to both CO_2 assimilation and leaf conductance was satisfactory.

MODELING SEASONAL TRENDS IN PHOTOSYNTHETIC BEHAVIOR

Photosynthetic properties of individual leaves change through time, both as a consequence of leaf aging and in response to a varying environment (e.g., acclimation to a changing light or temperature regime). In an attempt to discover if the model is capable of mimicking observed seasonal changes in leaf photosynthetic behavior, it was fit to several diurnal responses of assimilation and conductance measured on another Mediterranean sclerophyl, *Quercus coccifera* L., some of which have been previously published (Tenhunen et al., 1986, 1987). Note that no A vs. C_i response curves were available for model parameterization as described above; the simulations found in Fig. 2-10 and 2-11 are essentially fitting exercises in which P_{ml}, $V_{c_{max}}$, and the stomatal sensitivity factor, k, were varied to provide the best fit. Leaves measured on 16 May (Fig. 2-10, left) exhibited the highest rates of assimilation and conductance of the year. These were recently fully expanded leaves, soil water potentials were high, and temperatures were near optimal. Both $V_{c_{max}}$ and P_{ml} were high. As water stress developed through June and July, A and g_s both fell, and began to exhibit marked midday depression (Tenhunen et

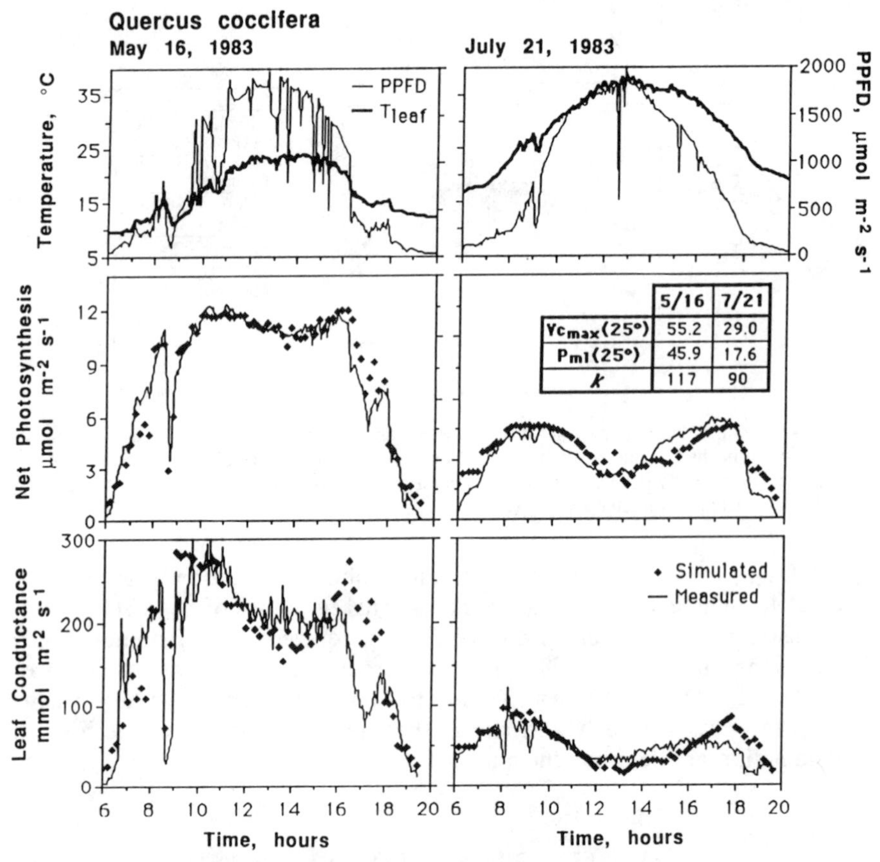

Fig. 2-10. Diurnal responses of net photosynthesis and leaf conductances of *Quercus coccifera*, measured on 16 May and 21 July 1983 in the field in Portugal. Also shown are diurnal courses of light (photosynthetic photon flux density [PPFD]) and leaf temperature. Symbols represent model simulations of the data. Values of P_{ml} and $V_{c_{max}}$ at 25 °C and the value of stomatal sensitivity (k) are shown.

al., 1985). Thus, in order to fit measured data, both $V_{c_{max}}$ and P_{ml} were reduced sharply (Fig. 2–10, right). Midday depression in the simulation is driven by a combination of superoptimal temperatures, which cause P_{ml} to decline, and extremely low relative humidities, which lead to a decrease in g_s and C_i.

With the onset of fall rains and cooler temperatures, assimilation rates recovered in October to near May values (data not shown) but, in the winter, rates were limited by cool temperatures and relatively low light (18 December, Fig. 2–11) despite relatively high photosynthetic capacity, as indicated by $V_{c_{max}}$ and P_{ml}. (Note that, although maximum leaf conductance was simulated well on this day, stomata failed to respond as rapidly as the model predicted to light-mediated fluctuations in assimilation; at cool temperatures, stomata of *Q. coccifera* were unresponsive to relatively short-term events; in one experiment, stomata required 3 h to reach equilibrium following a step change in irradiance [Tenhunen, 1984, unpublished data].) Another

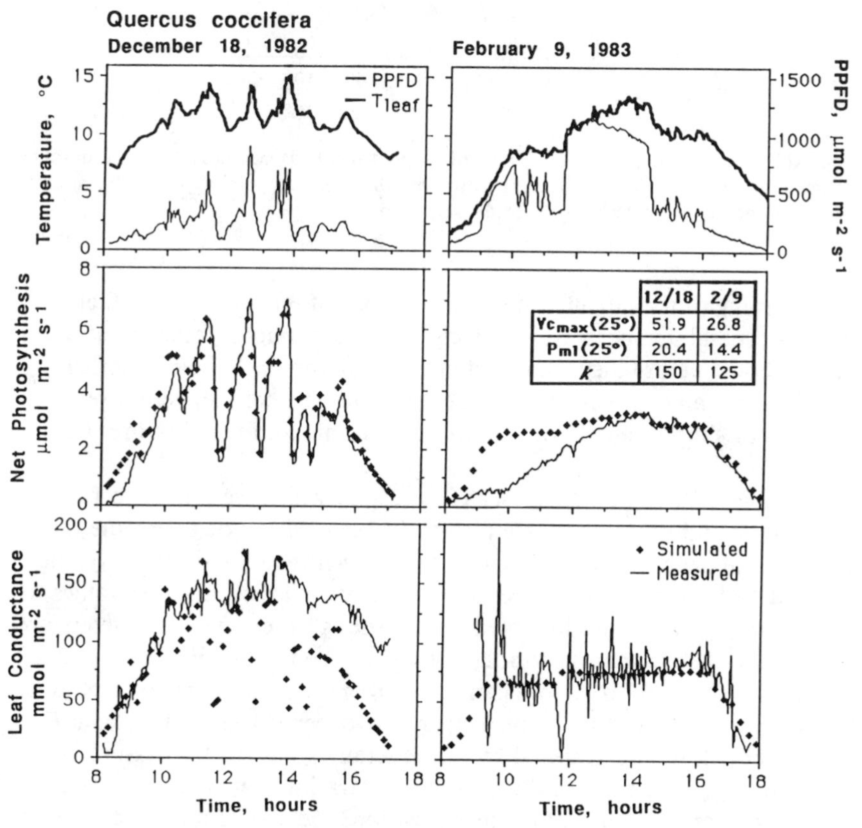

	12/18	2/9
Vcmax(25°)	51.9	26.8
Pml(25°)	20.4	14.4
k	150	125

Fig. 2–11. Diurnal responses of net photosynthesis and leaf conductances of *Quercus coccifera*, measured on 18 Dec. 1982 and 9 Feb. 1983 in the field in Portugal. Also shown are diurnal courses of light (photosynthetic photon flux density [PPFD]) and leaf temperature. Symbols represent model simulations of the data. Values of P_{ml} and $V_{c_{max}}$ at 25 °C and the value of stomatal sensitivity (k) are shown.

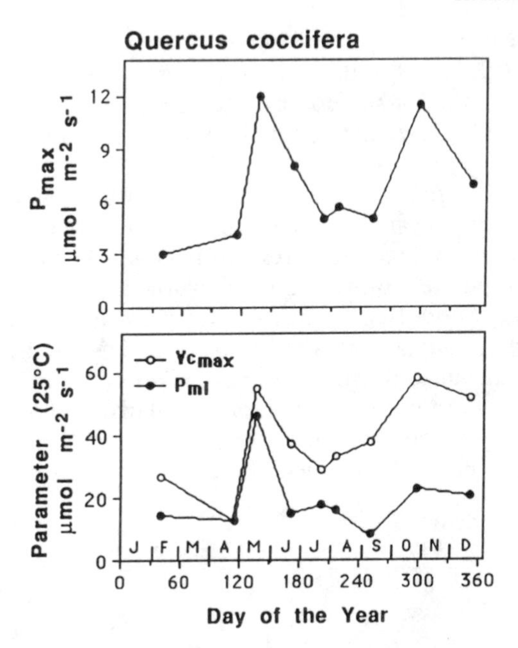

Fig. 2-12. Plot of (top) the maximum rates of net photosynthesis measured during the day nine times during the year (P_{max}), in comparison with (bottom) values of $V_{c_{max}}$ and P_{ml} at 25 °C used in model parameterizations for the same 9 d.

shortcoming of the model is apparent for data of 9 February, in which subzero overnight temperatures severely depressed photosynthetic rates for several hours before they recovered in early afternoon. Values of both $V_{c_{max}}$ and P_{ml} were among the lowest of the year, but whether they reflect seasonal patterns or simply short-term changes due to this cold-temperature stress is uncertain.

Values of $V_{c_{max}}$ and P_{ml}, scaled to 25 °C, which were used to fit data from these and other days, are plotted in Fig. 2-12, along with the maximum rate of net photosynthesis measured on each day. Not surprisingly, there is a marked correlation between measured assimilation and the values of $V_{c_{max}}$ and P_{ml}. It should be noted that, although these two parameters are the primary determinants of model behavior, the fits obtained in Fig. 2-10 and 2-11 also involved minor changes in other model parameters, including α and R_d, as well as shifts in the temperature dependencies of $V_{c_{max}}$ and P_{ml}.

Although we are gratified by the generally good fits to measured data found in Fig. 2-10 and 2-11, it is worth reiterating that these were not obtained using an independent parameterization, and should not be considered validation of either the model or the parameterization protocol. Nevertheless, they demonstrate that the model structure is capable of mimicking a wide range of leaf behavior, and that the model is driven largely by changes in two key parameters, $V_{c_{max}}$ and P_{ml}.

THE VALIDITY OF MODEL ASSUMPTIONS

The fundamental assumptions underlying this model of C_3 photosynthesis are (i) that photosynthesis may be explained on the basis of Rubisco activity, and (ii) that Rubisco activity is entirely limited either by RuBP concentration or, if RuBP is saturating, by the kinetic properties of the enzyme as modulated by the competition between CO_2 and O_2. If photosynthesis is equated with CO_2 assimilation (rather than with sucrose production, for example), then Assumption (i) seems justifiable on the basis that CO_2 uptake occurs at Rubisco. If the stoichiometry reflected in Eq. [1] is correct and 0.5 mol of CO_2 is liberated for every 1 mol of O_2 fixed, then modeling Rubisco activity alone is sufficient to closely approximate net CO_2 uptake (ignoring the small contribution of R_d). That Rubisco activity can, in fact, be accurately modeled, based on Assumption (ii), is more problematic.

The notion that photosynthesis at any given time is entirely limited by one factor to the exclusion of all others goes back to Blackman (1905), but the concept of an abrupt transition between RuBP-limited and RuBP-saturated conditions depicted in Fig. 2–1 and formalized in the min{} function of Eq. [6], although a useful modeling construct, is an oversimplification. The process of photosynthesis consists of a series of convergent metabolic pathways, and regulatory mechanisms exist to ensure a rough balance of activity between them (Woodrow & Berry, 1988; Sage et al., 1988). A few examples illustrate this point. At low light, where RuBP is assumed to be limiting in the model, Rubisco activity is strongly reduced (Perchorowicz & Jensen, 1983) and RuBP concentration remains higher than predicted. It is unclear in this instance whether Rubisco activity or RuBP concentration is truly limiting, and it is useful to think of the two as roughly co-limiting. Similarly, at high CO_2 where RuBP limits model behavior, Rubisco activity is strongly reduced (Sage et al., 1988). It is, therefore, clearly unjustifiable to use a model of this type to predict Rubisco activity or the size of metabolic pools. Model predictions of CO_2 flux, however, are identical whether RuBP is solely limiting or co-limiting with the Rubisco activation state. We suggest, therefore, that Assumption (ii) above is a necessary and approximately valid simplification enabling us to model an extremely complex system.

Another simplifying assumption, implicit in the use of C_i to drive the model, is that there is no resistance to CO_2 flux between the intercellular air space and the site of fixation, which is clearly not the case (Nobel, 1983). Evans et al. (1986), combining C isotope discrimination techniques with leaf gas exchange, estimated that the CO_2 concentration in the stroma could be reduced by as much as 60 $\mu L\ L^{-1}$ in rapidly photosynthesizing wheat (*Triticum aestivum* L.) leaves. As a result, our estimates of $V_{c_{max}}$, based on an overestimate of CO_2 concentration, must be considered maximum rates of carboxylation of the integrated leaf system, rather than of the enzyme per se. Since our goal is the prediction of leaf rather than enzyme behavior, this seems acceptable. It should be noted, however, that, although the drawdown in CO_2 may be significant, the concomitant drawdown in O_2 is trivial, resulting in an error in V_o/V_c and a slight overestimation of the photorespiratory contribution.

THE IMPORTANCE OF TRIOSE PHOSPHATE UTILIZATION

It has long been recognized that P_i may limit CO_2 uptake in isolated chloroplasts and that regulation via P_i is related to triose phosphate utilization and sink activity (Herold, 1980). It remains to be seen how significant P_i limitations are under natural conditions. Although previously observed only under saturating light and supranormal CO_2 levels (e.g., Sharkey, 1985b), it has been demonstrated recently that P_i limitations also occur under ambient CO_2 levels at relatively low but ecologically relevant temperatures (Sage & Sharkey, 1987; Labate & Leegood, 1988). Limitations arising from an imbalance between triose phosphate production and utilization are also implicated in cases where sink strength limits photosynthesis (Herold, 1980) and other situations accompanied by the accumulation of starch, including water stress (Sharkey, 1985b), prolonged exposure to high light (Azcon-Bieto, 1983), and growth at elevated CO_2 levels (DeLucia et al., 1985).

The effects of P_i are potentially important inasmuch as they provide the linkage between primary photochemical events and remote processes such as sucrose metabolism in the cytosol and sink strength of distant organs. Furthermore, P_i may mediate the effects of several kinds of stress. To the extent that P_i limitations can develop in the short term (hours to days), they represent a considerable parameterization challenge. We have attempted to incorporate P_i limitations into the model structure above. As additional evidence of P_i effects is reported (operating in the range of environmental conditions normally encountered), they can be better incorporated into the overall modeling effort.

CONCLUSION

In this chapter, we have described, in some detail, a physiologically based model of leaf photosynthesis. Any mechanistic model is limited by our understanding of the system in question and represents a compromise between mechanistic detail and practicality. By making several simplifying assumptions, we have realized a model requiring a manageable number of parameters (Table 2-1), approximately half of which may be considered constants. We have described the means of estimating values of the remaining parameters from gas-exchange measurements, which have become more or less routine with the development of field-portable photosynthesis systems.

Any model should be designed with specific goals in mind and we have used this model to elucidate the environmental controls on leaf CO_2 uptake and water loss. We have incorporated it into a canopy model (Caldwell et al., 1986) to help us understand the complex interactions of canopy architecture, leaf energy balance, and whole-canopy gas exchange. Although simple relative to the true complexity of leaf biochemistry, the model may well be too complex for inclusion into models addressing different questions, such as crop growth models, where more empirical models of gas exchange may be more practical. We believe that the considerable amount of biochemical knowledge incorporated into this leaf-level model, and its demonstrated effectiveness in mimicking leaf response, makes it a useful tool for developing and parameterizing phenomenological gas-exchange models for use in canopy and crop simulations.

ACKNOWLEDGMENT

This research was funded in part by a grant from the Carbon Dioxide Research Division of the U.S. Dep. of Energy, Grant no. DE-FG03-86ER60490.

REFERENCES

Azcon-Bieto, J. 1983. Inhibition of photosynthesis by carbohydrates in wheat leaves. Plant Physiol. 173:681–686.

Badger, M.R., and G.J. Collatz. 1977. Studies on the kinetic mechanism of ribulose-1,5-bisphosphate carboxylase and oxygenase reactions, with particular reference to the effect of temperature on kinetic parameters. Year Book Carnegie Inst. Washington 76:355–361.

Ball, J.T., I.E. Woodrow, and J.A. Berry. 1987. A model predicting stomatal conductance and its contribution to the control of photosynthesis under different environmental conditions. p. 221–224. *In* J. Biggins (ed.) Prog. Photosynth. Res. Proc. Int. Congr. 7th, Providence. 10–15 Aug. 1986. Vol. 4. Kluwer, Boston.

Berry, J., and O. Björkman. 1980. Photosynthetic temperature response and adaptation to temperature in higher plants. Annu. Rev. Plant Physiol. 31:491–543.

Blackman, F.F. 1905. Optima and limiting factors. Ann. Bot. (London) 19:281–295.

Brooks, A., and G.D. Farquhar. 1985. Effect of temperature on the CO_2/O_2 specificity of ribulose-1,5-bisphosphate carboxylase/oxygenase and the rate of respiration in the light. Estimates from gas-exchange measurements on spinach. Planta 165:397–406.

Caldwell, M.M., H.-P. Meister, J.D. Tenhunen, and O.L. Lange. 1986. Canopy structure, light microclimate and leaf gas exchange of *Quercus coccifera* L. in a Portuguese macchia: Measurements in different canopy layers and simulations with a canopy model. Trees (Berlin) 1:25–41.

DeLucia, E.H., T.W. Sasek, and B.R. Strain. 1985. Photosynthetic inhibition after long-term exposure to elevated levels of atmospheric carbon dioxide. Photosynth. Res. 7:175–184.

Ehleringer, J., and O. Björkman. 1977. Quantum yields for CO_2 uptake in C_3 and C_4 plants. Dependence on temperature, CO_2 and O_2 concentration. Plant Physiol. 59:86–90.

Evans, J.R., and J.R. Seemann. 1984. Differences between wheat genotypes in specific activity of ribulose-1,5-bisphosphate carboxylase and the relationship to photosynthesis. Plant Physiol. 74:759–765.

Evans, J.R., T.D. Sharkey, J.A. Berry, and G.D. Farquhar. 1986. Carbon isotope discrimination measured concurrently with gas exchange to investigate CO_2 diffusion in leaves of higher plants. Aust. J. Plant Physiol. 13:281–292.

Farquhar, G.D. 1979. Models describing the kinetics of ribulose bisphosphate carboxylase/oxygenase. Arch. Biochem. Biophys. 193:456–468.

Farquhar, G.D., and S. von Caemmerer. 1982. Modelling of photosynthetic response to environment. p. 549–587. *In* O.L. Lange et al. (ed.) Encyclopedia of plant physiology. New Ser. Vol. 12B. Physiological Plant Ecology II. Springer-Verlag, Berlin.

Farquhar, G.D., S. von Caemmerer, and J.A. Berry. 1980. A biochemical model of photosynthetic CO_2 assimilation in leaves of C_3 species. Planta 149:78–90.

Farquhar, G.D., and S.C. Wong. 1984. An empirical model of stomatal conductance. Aust. J. Plant Physiol. 11:191–210.

Hall, N.P., and A.J. Keys. 1983. Temperature dependence of the enzymic carboxylation and oxygenation of ribulose 1,5-bisphosphate in relation to effects of temperature on photosynthesis. Plant Physiol. 72:945–948.

Harley, P.C., J.D. Tenhunen, and O.L. Lange. 1986. Use of an analytical model to study limitations to net photosynthesis in *Arbutus unedo* under field conditions. Oecologia 70:393–401.

Harley, P.C., J.A. Weber, and D.M. Gates. 1985. Interactive effects of light, leaf temperature, carbon dioxide and oxygen on photosynthesis in soybean. Planta 165:249–263.

Herold, A. 1980. Regulation of photosynthesis by sink activity—The missing link. New Phytol. 86:131–144.

Hodgman, C.D., R.C. Weast, and S.M. Selby. 1958. Solubility of gases in water. p. 1706–1707. *In* C.D. Hodgman (ed.) Handbook of chemistry and physics. 40th ed. CRC Press, Boca Raton, FL.

Johnson, F., H. Eyring, and R. Williams. 1942. The nature of enzyme inhibitions in bacterial luminescence: Sulfanilamide, urethane, temperature, and pressure. J. Cell Comp. Physiol. 20:247–268.

Jordan, D.B., and W.L. Ogren. 1984. The CO_2/O_2 specificity of ribulose 1,5-bisphosphate carboxylase/oxygenase. Dependence on ribulose-bisphosphate concentration, pH and temperature. Planta 161:308–313.

Keys, A.J. 1986. Rubisco: Its role in photorespiration. Proc. R. Soc. London B 313:325–336.

Kirschbaum, M.U.F., and G.D. Farquhar. 1984. Temperature dependence of whole-leaf photosynthesis in *Eucalyptus pauciflora* Sieb. ex Spreng. Aust. J. Plant Physiol. 11:519–538.

Ku, S.-B., and G.E. Edwards. 1977. Oxygen inhibition of photosynthesis. II. Kinetic characteristics as affected by temperature. Plant Physiol. 59:986–990.

Labate, C.A., and R.C. Leegood. 1988. Limitation of photosynthesis by changes in temperature. Factors affecting the response of carbon-dioxide assimilation to temperature in barley leaves. Planta 173:519–527.

Mohanty, P., and J.S. Boyer. 1976. Chloroplast response to low leaf water potentials. IV. Quantum yield is reduced. Plant Physiol. 57:704–709.

Nobel, P.S. 1983. Biophysical plant physiology and ecology. W.H. Freeman and Co., San Francisco.

Perchorowicz, J.T., and R.G. Jensen. 1983. Photosynthesis and activation of ribulose bisphosphate carboxylase in wheat seedlings. Plant Physiol. 71:955–960.

Powles, S.B. 1984. Photoinhibition of photosynthesis induced by visible light. Annu. Rev. Plant Physiol. 35:15–44.

Sage, R.F., and T.D. Sharkey. 1987. The effect of temperature on the occurrence of O_2 and CO_2 insensitive photosynthesis in field grown plants. Plant Physiol. 84:658–664.

Sage, R.F., T.D. Sharkey, and J.R. Seemann. 1988. The in-vivo response of the ribulose-1,5-bisphosphate carboxylase activation state and the pool sizes of photosynthetic metabolites to elevated CO_2 in *Phaseolus vulgaris* L. Planta 174:407–416.

Sharkey, T.D. 1985a. Photosynthesis in intact leaves of C_3 plants: Physics, physiology and rate limitations. Bot. Rev. 51:53–105.

Sharkey, T.D. 1985b. O_2-insensitive photosynthesis in C_3 plants. Its occurrence and a possible explanation. Plant Physiol. 78:71–75.

Smith, E. 1937. The influence of light and carbon dioxide on photosynthesis. Gen. Physiol. 20:807–830.

Tenhunen, J.D., W. Beyschlag, O.L. Lange, and P.C. Harley. 1987. Changes during summer drought in leaf CO_2 uptake rates of macchia shrubs growing in Portugal. p. 305–327. *In* J.D. Tenhunen et al. (ed.) Plant response to stress—Functional analysis in Mediterranean ecosystems. Springer-Verlag, Heidelberg, Germany.

Tenhunen, J.D., O.L. Lange, P.C. Harley, W. Beyschlag, and A. Meyer. 1985. Limitations due to water stress on leaf net photosynthesis of *Quercus coccifera* in the Portuguese evergreen scrub. Oecologia 67:23–30.

Tenhunen, J.D., J.A. Weber, C.S. Yocum, and D.M. Gates. 1976. Development of a photosynthesis model with an emphasis on ecological applications. II. Analysis of a data set describing the P_m surface. Oecologia 26:101–109.

von Caemmerer, S., and G.D. Farquhar. 1984. Effects of partial defoliation, changes of irradiance during growth, short-term water stress and growth at enhanced $p(CO_2)$ on the photosynthetic capacity of leaves of *Phaseolus vulgaris* L. Planta 160:320–329.

Walker, D.A., and A. Herold. 1977. Can the chloroplast support photosynthesis unaided? Plant Cell Physiol. (spec. iss.) 18:1–7.

Wong, S.C., I.R. Cowan, and G.D. Farquhar. 1979. Stomatal conductance correlates with photosynthetic capacity. Nature (London) 282:424–426.

Woodrow, I.E., and J.A. Berry. 1988. Enzymatic regulation of photosynthetic CO_2 fixation in C_3 plants. Annu. Rev. Plant Physiol. Plant Mol. Biol. 39:533–594.

3 Modeling Canopy Photosynthetic Response to Carbon Dioxide, Light Interception, Temperature, and Leaf Traits

Basil Acock

USDA-ARS
Systems Research Lab
Beltsville, MD

When I began my research career 24 yr ago at the Glasshouse Crops Research Institute (GCRI), Littlehampton, England, the agricultural engineers had just given some new toys to glasshouse growers. These were controllers for temperature and CO_2 concentration based on measurements of light, temperature, and humidity. Glasshouse operators, having purchased these new controllers, asked the plant physiologists how to set the control knobs. We did not, and still do not, know the answer to that question. However, it was one of the factors that prompted the initiation of a modeling program at the Institute. My role as an experimenter involved collecting data on canopy photosynthesis and the growth of glasshouse crops in daylit controlled-environment cabinets. John Thornley and David Charles-Edwards were the modelers in the group and John Warren Wilson was the leader. Publications by the group often listed authors alphabetically, which gave me more credit than I deserve. This chapter describes a canopy photosynthesis model begun by that group, further developed by me, and incorporated in the soybean crop simulator GLYCIM (Acock et al., 1985b).

MODELING PHILOSOPHY

The GCRI group made many measurements of leaf and canopy photosynthesis under various conditions and developed several detailed leaf photosynthesis models. When it came time to model canopy photosynthesis, however, the unanimous decision was to start with an empirical hyperbolic curve to describe leaf photosynthesis. There were, even then, many good sets of data describing leaf photosynthesis as a function of light flux density, CO_2 concentration or both. The problem that we faced was to go from the

 Modeling Crop Photosynthesis—from Biochemistry to Canopy. CSSA Special Publication no. 19.

leaf level to the canopy level. Several researchers, including Davidson and Philip (1958) and Verhagen et al. (1963), had tried to integrate leaf photosynthesis throughout the entire canopy to predict canopy photosynthesis and had failed in their attempts. The prevailing view at the time was that such integration was not possible by either analytical or numerical means because the pattern of variation in leaf photosynthetic characteristics within the canopy was not understood. We, therefore, decided to use simple equations that gave photosynthesis a hyperbolic dependence on both light and CO_2 for both single-leaf and whole-canopy photosynthesis. Then we sought to relate the parameters in the leaf-level model to those in the canopy-level model using analytical integration. The rationale was that a model with this structure could be parameterized either from canopy photosynthetic or leaf photosynthetic data.

FITTING RECTANGULAR HYPERBOLAE TO LEAF PHOTOSYNTHESIS CURVES

The rectangular hyperbola has been used by many researchers to describe leaf photosynthesis curves. The observed curves often appear to bend more sharply than a rectangular hyperbola and, as a result, some researchers have gone on to use a nonrectangular hyperbola (e.g., Johnson & Thornley, 1984). This introduces an extra parameter. One form of the equation giving a hyperbolic dependence on leaf net and gross photosynthetic rates (P_n and P_g, respectively) on both the light flux density at the leaf surface (I) and CO_2 concentration (C) is

$$P_n = \alpha I \tau C/(\alpha I + \tau C) - R \quad [1]$$

and

$$P_g = P_n + R$$
$$= \alpha I \tau C/(\alpha I + \tau C) \quad [2]$$

where α is leaf light-utilization efficiency (quantum efficiency), τ is leaf conductance to CO_2 transfer, and R is leaf respiration rate in the absence of photosynthesis. Photorespiration is ignored in these equations, but will be introduced below. For the derivation of the equation, examples of various forms it can take, references to its early use, and an example of how well it fits data for *Amaranthus caudatus* L. (a C_4 plant), see Acock et al. (1971). An example of the equation fitted to data for soybean [*Glycine max* (L.) Merr., a C_3 plant] is given in Acock and Allen (1985).

The significance of the parameters in the equation can best be understood by referring to Fig. 3–1. If we plot P_n against I, the intercept of the curve at zero light is R, the initial slope is α, and the asymptote toward which the

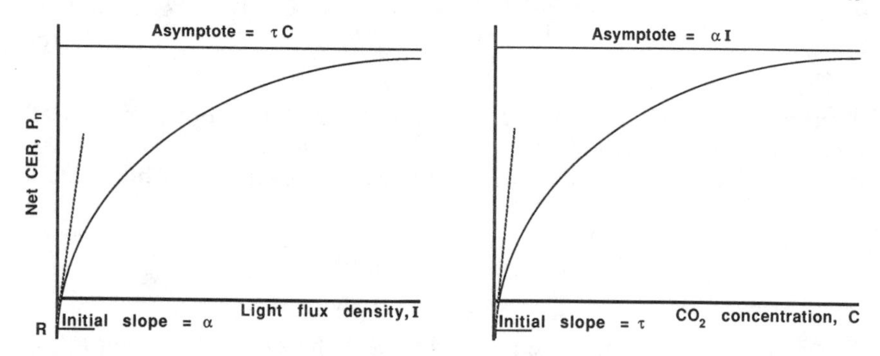

Fig. 3-1. Diagram showing the relationship between the variables and parameters in Eq. [1] and [2].

curve tends is τC. If we plot P_n as a function of C, the intercept is still R, the initial slope is τ, and the asymptote is αI.

The term R includes both growth and maintenance respiration. Since these sources of respiration will be modeled separately below, we will develop the model for P_g first. When the term P_g is defined as $P_n + R$, it does not take account of photorespiration. As a result, α derived from Eq. [1] using data on P_n will be too small to give a true P_g value when used in Eq. [2].

Some authors, e.g., Charles-Edwards (1978), have included photorespiration explicitly in Eq. [1] by writing

$$P_n = \alpha_m I(\tau C + \beta O)/(\alpha_m I + \tau C) - R \qquad [3]$$

where α_m is α in the absence of O_2, O is O_2 concentration, and β is leaf conductance to O_2 transfer. Photorespiration rate (R_l) is then given by

$$R_l = \alpha_m I \beta O/(\alpha_m I + \tau C) \qquad [4]$$

and the true P_g accounting for photorespiration (P_g')

$$P_g' = P_n + R + R_l = \alpha_m I \tau C/(\alpha_m I + \tau C). \qquad [5]$$

Thus Eq. [2] can easily be corrected for photorespiration by substituting α_m for α, and we will see below how α and α_m are related.

LIGHT INTERCEPTION BY A CLOSED CANOPY

The equation most commonly used to describe light interception by the canopy is often referred to as Beer's law (Acock et al., 1970). The same law is sometimes attributed to Lambert or Bouguer. It was originally developed to describe light attentuation in a suspension of small particles, such as a fog or muddy water, but it has also been applied successfully to crop canopies in the form

$$I' = I_o \exp(-KL) \qquad [6]$$

where I' is light flux density at a given position in the canopy below a leaf area index of L, I_o is light flux density above the canopy, and K is the canopy extinction coefficient. Taking the natural logarithm of both sides, we get

$$\ln(I') = \ln(I_o) - KL. \qquad [7]$$

This is clearly the equation for a straight line with slope K. If we plot $\ln(I')$ against L (Fig. 3-2), we find that K is a function of the orientation of the leaves forming the canopy. Plants like grasses, with narrow erect leaves, have a small value of K and plants like cotton, with broad horizontal leaves, have a large value of K. In other words, light penetrates farther into a grass canopy than into a cotton canopy.

Beer's law, strictly, only applies to small, randomly oriented particles (leaves) and L is a measure of the number of particles passed by the light beam. A light beam from the sun at noon in summer, when it is virtually overhead, passes far fewer leaves on its way to the soil than a beam of sunlight in late afternoon. If Beer's law is naively applied to crop canopies without taking into account this change in effective leaf area index, K will appear to vary with the time of day. Even making this allowance, K varies with solar altitude for some crops in certain stages of growth. Many crops have leaves that are small and randomly oriented, however, and for these the parameter K can be thought of as a crop characteristic that is virtually constant.

Now, Beer's law predicts the downward flux density of light past a certain point in the canopy but takes no account of light scattering, absorption, and transmission by the leaves. To accommodate these phenomena, Saeki (1960, 1963) proposed the extension to the Beer's law:

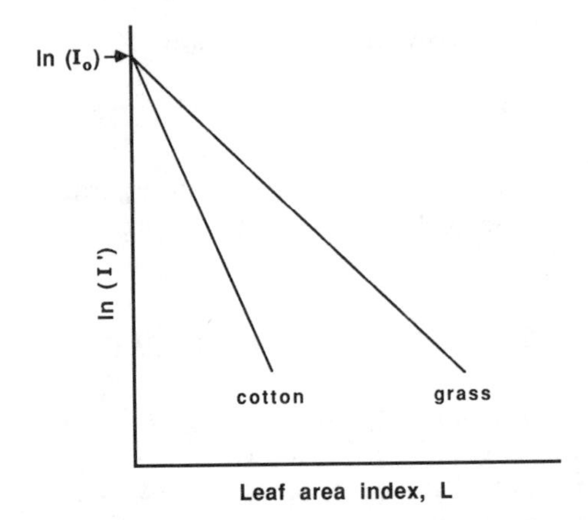

Fig. 3-2. Diagram illustrating Beer's law of light interception applied to a crop canopy, where I_o is light flux density above the canopy and I' is light flux density within the canopy under a leaf area index of L.

$$I = I_o \exp(-KL)\, K/(1 - m) \qquad [8]$$

where m equals the proportion of light transmitted by the leaf. For the derivation of this equation, see Acock et al. (1978). The new parameter, m, is easily measured, varies only slightly from leaf to leaf within the canopy, and can also be treated as a crop characteristic.

LIGHT INTERCEPTION BY CROP ROWS

It is easy to see how the concept of leaf area index can be applied to a closed canopy but, for some of the growing season, many agronomic crops exist in rows with bare soil in between. What is the appropriate term to use in place of leaf area index for crops in this stage of development?

For modeling purposes, we can approximate the volume of space occupied by a crop row with a regular three-dimensional figure. Depending on the crop, this might be circular, elliptical, rectangular, or triangular in cross section. Knowing the dimensions of this figure, the orientation of the row, the latitude, day of the year, and time of day, it is a simple exercise in geometry to calculate the length of the shadow. The relationship between the true solar elevation (χ') and the apparent solar elevation at right angles to the row (χ) is given by

$$\tan(\chi) = \tan(\chi')/\sin\delta \qquad [9]$$

where δ is the angle between solar azimuth and row orientation. Equations to calculate solar elevation and solar azimuth from latitude, day of the year, and time of day are given in most books of celestial mechanics (e.g., Iqbal, 1983; List, 1963). Taking as an example a crop row with approximately circular cross section, the shadow length measured at right angles to the row (D) is given by

$$D = Z/\sin\chi \qquad [10]$$

where Z is the height of the crop row. This shadowed area is the soil area effectively occupied by the crop for purposes of light interception and the effective leaf area index (L_e) to use in our equations is

$$L_e = \text{crop leaf area/shadow area.} \qquad [11]$$

In the example of the row with circular cross section,

$$L_e = LZ/D$$
$$= L \sin\chi. \qquad [12]$$

However, Eq. [11] is more general and works with all shapes of row cross section. If shadow length is greater than row spacing, the whole soil is in shadow and it is best to treat the crop as a closed canopy. For a closed canopy,

$$L_e = L \tan\chi \quad [13]$$

Comparing Eq. [12] and [13], it can be seen that the error involved in treating a row crop as a closed canopy at small values of χ is negligible. This method of calculating L_e assumes that the leaves are distributed uniformly over the shadowed area of soil. This is not true, of course, but more exact calculations assuming the leaves are distributed uniformly within the crop volume show that the error introduced is small and light interception is virtually constant across the entire width of a crop row.

INTERCEPTION OF DIFFUSE LIGHT

The same technique can be used for estimating L_e and the interception of diffuse light by crop rows. This is done by integrating light coming from various parts of the sky. If we assume a uniform overcast sky, it is usually sufficient to perform the calculation for light coming from 10 positions between 0 and 90° elevation measured at right angles to the row. For a standard overcast sky, it would be necessary to integrate across the full hemisphere of sky. Analytical solutions to these integration problems should be possible, and may have been developed by others.

VARIATION IN LEAF PHOTOSYNTHETIC PARAMETERS WITHIN THE CANOPY

Values of α measured at ambient CO_2 concentration vary slightly from one laboratory to another, possibly because of differences in the light sources used to obtain them. However, the results show that α is virtually constant across all C_3 species when measured at the same temperature and CO_2 concentration (Charles-Edwards, 1978; Björkman & Demmig, 1987). There is no evidence to suggest that α varies with leaf position in the canopy, either. We can, therefore, treat α as a constant for all leaves on C_3 crops growing in a given environment. Below, we will see how α varies with temperature and the concentrations of CO_2 and O_2.

The values of τ and R both decrease with depth in the canopy. These decreases are not related to leaf age but are rather a function of the amount of light reaching the leaf. Similarly, τ and R measured on the uppermost fully exposed leaves of plants decrease as the light flux density in which they are grown decreases (Charles-Edwards & Ludwig, 1975). In fact, τ and R for plants grown in different light flux densities and those for leaves shaded within the canopy all show the same hyperbolic dependence on light flux (Hozumi et al., 1972; Kira, 1975). Acock et al. (1978) found that, for a tomato canopy,

$$\tau = aS/(1 + bS) \quad [14]$$

$$R = eS/(1 + fS) \quad [15]$$

where S is the average light flux density recently experienced by the leaf, and a, b, e, and f are parameters. The use of the word *recently* is deliberately vague, because we do not know how rapidly leaves adjust to a change in light flux density. Bunce et al. (1977) have shown that many physiological changes occur within one day, but anatomical changes in response to a change in light flux density take longer. In our work, we average light flux density during the past week and use it in Eq. [14] to calculate τ. We know that R is more responsive to the environment than Eq. [15] suggests, however, so to calculate R we use another method that will be discussed below.

When S was measured in W m^{-2}, τ in m s^{-1}, and R in mg CO_2 m^{-2} s^{-1}, the parameters were found to have the following values for a tomato canopy (Acock et al., 1978):

$$a = 8.5 \times 10^{-5} \text{ m}^3 \text{ J}^{-1}$$
$$b = 2.1 \times 10^{-2} \text{ m}^2 \text{ s J}^{-1}$$
$$e = 2.4 \times 10^{-3} \text{ mg CO}_2 \text{ J}^{-1}$$
$$f = 1.9 \times 10^{-2} \text{ m}^2 \text{ s J}^{-1}$$

By analogy with Eq. [8],

$$S = S_o \exp(-KL)K/(1 - m) \quad [16]$$

where S_o is average light flux density recently incident at the top of the canopy.

RELATING SINGLE-LEAF PHOTOSYNTHETIC CHARACTERISTICS TO CANOPY CHARACTERISTICS

Combining Eq. [2], [8], [14], and [16] and integrating across the canopy, we get

$$P_{gc} = (aC/bK) \ln\{[b\alpha I_o S_o K + (1 - m)(\alpha I_o + aS_oC)/ [b\alpha I_o S_o K \exp(-KL) + (1 - m)(\alpha I_o + aS_oC)]\} \quad [17]$$

where P_{gc} is canopy gross photosynthetic rate per unit ground area. See Acock et al. (1978) for intermediate steps in the derivation.

Now, by analogy with Eq. [2], we can write

$$P_{gc} = \alpha_c I \tau_c C/(\alpha_c I + \tau_c C) \quad [18]$$

where α_c is canopy light utilization efficiency and τ_c is canopy conductance to CO_2 transfer. From Fig. 3–1, we can see that $\alpha = P_g/I = (P_n + R)/I$

as I approaches 0. Therefore, starting with Eq. [17], if we differentiate P_{gc} with respect to I_o and take the limit as $I_o \rightarrow 0$,

$$\alpha_c = dP_{gc}/dI_o \text{ as } I_o \rightarrow 0$$
$$= \alpha[1 - \exp(-KL)/(1 - m)] \quad [19]$$

Similarly, from Fig. 3-1, $\tau C = P_g$ at $I_o = \infty$. Therefore, taking the limit of P_{gc} as $I_o \rightarrow \infty$,

$$\tau_c C = \lim(P_{gc}) \text{ as } I_o \rightarrow \infty$$
$$= aC/bK \ln\{[bS_oK + (1 - m)]/$$
$$[bS_oK \exp(-KL) + (1 - m)]\} \quad [20]$$

Dividing by C,

$$\tau_c = a/bK \ln\{[bS_oK + (1 - m)]/[bS_o K \exp(-KL) + (1 - m)]\}. \quad [21]$$

Equations [19] and [21] enable us to calculate the canopy photosynthetic characteristics α_c and τ_c from the more stable leaf photosynthetic characteristics. These canopy characteristics can then be used in Eq. [18] to calculate P_{gc}.

EFFECT OF TEMPERATURE ON PHOTOSYNTHESIS

Most data on the effect of temperature on photosynthesis show a gradual increase in photosynthesis as temperature increases, a plateau of little or no response, followed by a decrease in photosynthesis at high temperatures (e.g., Osmond et al., 1980; Jurik et al., 1984). It is not generally recognized that the conditions under which the measurements are made have a profound influence on the shape of the relationship obtained.

The most comprehensive set of data on photosynthesis and temperature known to me has never previously been published. The data are briefly mentioned in Ludwig and Withers (1978), and were obtained when they were working with the team at GCRI. The data (Fig. 3-3) clearly show that the shape of the photosynthetic response to temperature depends on the environmental conditions under which it was measured. The curves are all separate at 15° C and above, but run together below that. The curves demonstrate that temperature acts as an additional limiting factor—additional, that is, to light and CO_2. But how can we find out what the temperature limitation looks like in the absence of other limitations? To determine this, we would have to measure photosynthesis at a range of temperatures in very high light and high CO_2 concentration.

Hofstra and Hesketh (1969) did something like this for several species when they measured photosynthesis in O_2-free air at various temperatures.

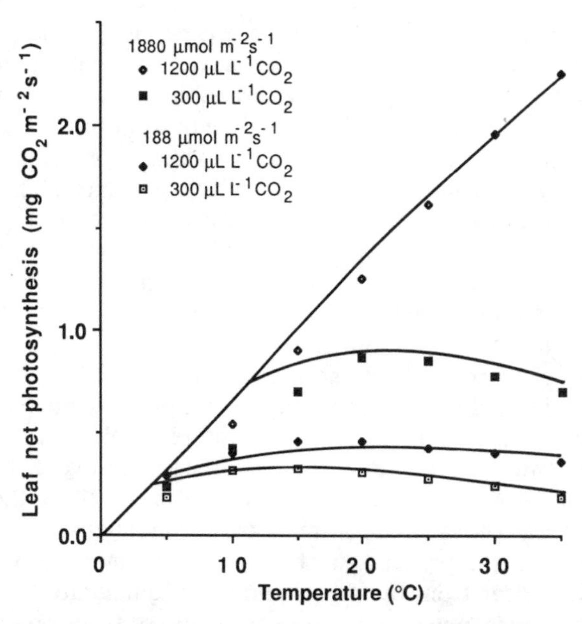

Fig. 3–.3 Leaf net photosynthetic rates measured at various temperatures, light flux densities, and CO_2 concentrations on tomato plants grown at 20° C and 380 μmol photon m^{-2} s^{-1} with a 16-h photoperiod (Ludwig & Withers, 1978). The points are measured data and the curves are predicted using a simple model described in the text.

Their data on gross photosynthesis for soybean leaves are shown in Fig. 3–4. Their data can be fitted very nicely with two straight lines and the datum point near the apex of the triangle, suggesting that very little rounding occurs

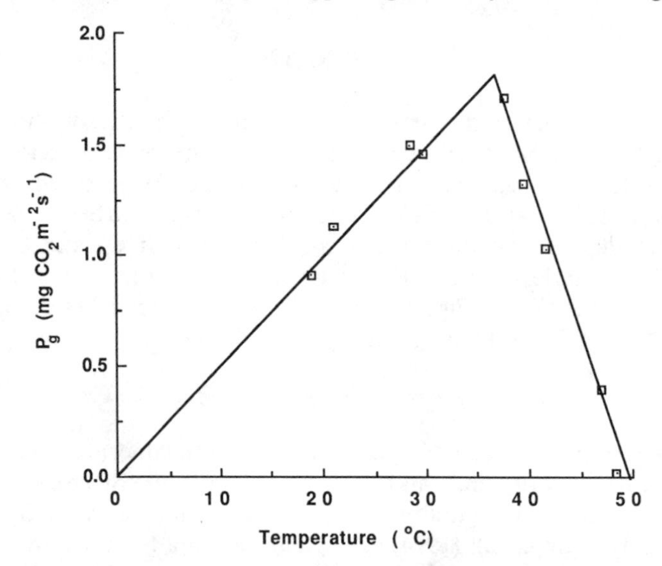

Fig. 3–4. Relationship between temperature and soybean leaf gross photosynthetic rate (P_g) in O_2-free air (Hofstra & Hesketh, 1969).

at the apex. The triangle in Fig. 3–4 defines the permissible range of values for photosynthetic rate for a given temperature. The peak photosynthetic rate in Fig. 3–4 is rather low, however, probably because the plants were raised in low light. To make practical use of the relationship, we assume that the general shape holds and we scale the apex value appropriately. Having done this scaling, we can use Fig. 3–4 to calculate a maximum photosynthetic rate as limited by temperature. Then we can use our equations relating photosynthesis to C and I to calculate a maximum photosynthetic rate as limited by C and I. Comparing the two values of photosynthetic rate, we accept the lowest value as being the value limited by C, I, or temperature.

Using this technique, we attempted to simulate Ludwig and Wither's photosynthesis data. For leaf net photosynthetic rate as a function of C and I, we used Eq. [3] and made respiration a function of temperature with a Q_{10} of 2.0, i.e., for every 10° C rise in temperature, respiration rate doubles. Combining this with a temperature function like that in Fig. 3–4 but scaled appropriately at the apex we had a model that explained 92% of the variation in the original data. The simulations are shown as continuous lines in Fig. 3–3.

A leaf model of this type can only be applied to a canopy using numerical integration. It is clear from the model and the original data that leaves in the highest light environment, i.e., those at the top of the canopy, are most likely to be temperature limited. Therefore, in applying this model to canopy photosynthesis, I start with the uppermost leaf layer (L =0.5) and determine if temperature is limiting. I then move progressively down through the canopy (L = 1.5, 2.5, etc.) until a layer is found where temperature is no longer limiting. Quite often, temperature does not even limit the photosynthetic response of the uppermost fully exposed leaves.

RESPIRATION

Above, we deferred a detailed discussion of photorespiration. One of the effects of increased CO_2 concentration is to decrease photorespiration. Ribulose-1,5-bisphosphate carboxylase-oxygenase (Rubisco) can act as both a carboxylase and an oxygenase. The extent to which it behaves as one or the other depends on the relative concentrations of O_2 and CO_2 in the choroplast. Charles-Edwards (1978) derived Eq. [3] from a consideration of enzyme kinetics. From Fig. 3–1, $\alpha = P_n/I$ as I approaches 0. Therefore, differentiating Eq. [3] with respect to I, we obtain

$$\alpha = \alpha_m (1 - \beta O/\tau C). \qquad [22]$$

Like α for a given temperature and CO_2 concentration, α_m is a conservative quantity and can be taken as a constant. But we now have to evaluate an additional parameter, β, the leaf conductance to O_2 transfer. Fortunately, Laing et al. (1974) have published data for soybean leaves that enable us to evaluate the ratio β/τ. This turns out to be a function of temperature:

$$\beta/\tau = 1.2 \times 10^{-4} \exp(0.0295T) \quad [23]$$

where T is the temperature at the chloroplast in ° C. We now have enough information to calculate α_m from α and we can use Eq. [5] to calculate P_g' in our model.

Equations [4] and [5] follow from Eq. [3] and, from them, the ratio of R_l to P_g' is given by

$$R_l/P_g' = \alpha_m I\beta O/(\alpha_m I + \tau C)(\alpha_m I + \tau C)/\alpha_m I\tau C$$
$$= \beta O/\tau C \quad [24]$$

therefore,

$$R_l = P_g'(\beta O/\tau C). \quad [25]$$

Equation [25] explicitly allows for the response of photorespiration to temperature and the concentrations of O_2 and CO_2.

In common with most models, maintenance respiration is handled according to the estimates of Penning de Vries (1975), who estimated that leaf protein turnover costs between 28 and 53 mg glucose g^{-1} protein d^{-1} and that ion leakage costs 8 to 10 mg glucose g^{-1} dry weight d^{-1} (temperature unspecified). If these figures for maintenance respiration are applied to all of the tissues in the plant, they rapidly become a major sink for C in the mature plant and cause unrealistic reductions in crop yield. Some cells, such as xylem elements, do not require maintenance and it may be that other tissues, such as the root cortex, are not maintained by the plant but are allowed to deteriorate. The tissues that will almost certainly be maintained by the plant are the leaf laminae. In the soybean crop simulator GLYCIM (Acock et al., 1985b), we therefore assume that only the leaf laminae are maintained and that the maintenance respiration has a Q_{10} of 2.0.

Growth respiration is not considered a part of the photosynthetic process. Instead, it is charged as the cost of synthesizing new tissues in other parts of the plant.

LEAF NITROGEN, SENESENCE, AND WATER RELATIONS

The response of leaf photosynthesis to N content has been put into our model as an empirical function (f_N). The relationship we use is

$$f_N = (1.65N - 3.3)/(1.32N - 1.64) \quad [26]$$

where N is leaf N concentration. This was obtained by fitting a curve to data of Boote et al. (1978). Almost certainly, leaf N concentration should be treated as another limiting factor on photosynthesis, but we have insufficient data to allow us to do this. The function f_N is therefore used as a multiplier, even though we know this is not a satisfactory way to interact factors.

As most canopies develop, old leaves become covered by successive layers of young leaves. As we have seen already, this reduction in the light level to which they are exposed reduces their photosynthetic rate. However, even leaves that remain fully exposed at the top of the canopy gradually senesce and lose their ability to photosynthesize. In a growing canopy, the reduction in photosynthetic capacity caused by older leaves becoming buried is far greater than the reduction caused by leaf senescence. But when a determinant crop canopy reaches the end of the vegetative phase and the uppermost fully expanded leaves remain fully exposed, then it is necessary to apply a senescence factor (f_s). For the soybean crop simulator GLYCIM, we have done this empirically using a multiplication factor derived from data of Woodward (1976):

$$\begin{aligned} &\text{for } A < 10,\ f_s = 1.0 \\ &\text{for } 10 < A < 60,\ f_s = \sin(72.0 + 1.8A) \\ &\text{for } A > 60,\ f_s = 0.0 \end{aligned} \qquad [27]$$

where A is the age of the leaf in days after full expansion. Senescing leaves as a function of age is unsatisfactory, but there is too little information about the phenomenon for us to model the processes involved.

In addition to normal aging, leaves senesce because (i) they are heavily shaded and are no longer self-supporting on a daily basis, (ii) the whole plant is N deficient and is moving N from older to younger leaves, and (iii) water stress is accelerating senescence. These factors are extrinsic to the canopy photosynthesis module in most whole-plant models, but they also cause leaves to senesce in the canopy and are important in predicting canopy photosynthesis over a whole growing season.

Effects of plant water relations on photosynthesis through stomatal conductance are also estimated in other parts of whole-plant models and will not be discussed here. In GLYCIM, photosynthetic rate is reduced when potential transpiration rate (E_0) exceeds actual transpiration rate (E). Photosynthetic rate is made proportional to E/E_0 when this occurs. Actual transpiration rate, E, is the maximum rate at which the plant roots can take up water in the prevailing conditions, or E_0, whichever is smaller.

SINK STRENGTH

Using data derived from measurements on tomato leaves, we assume that 35% of the current net photosynthate is immediately translocated. The rest is placed in the starch storage pool, from where it is continuously remobilized at the rate of 0.125 of the pool per hour (Ho, 1978).

It has been shown for many crops that reducing the sink strength can also suppress the photosynthetic rate (e.g., Clough et al., 1981). In these experiments, the source/sink ratio is usually manipulated in some way, but there is no reason to suppose that sink strength may not limit at some times during normal crop growth. For instance, plants grown in high light and low temperatures have slow growth rates, high starch concentrations, and low photosynthetic rates. Sink strength is limiting.

In a recent experiment, Acock et al. (1985b) followed the photosynthetic rate of a closed soybean canopy every day for an entire growing season. The results are plotted in Fig. 3–5. On Day 25, the L was 1.0 but, by Day 35, L was 3.0 and virtually all of the light was being intercepted. This increase in L would account for a doubling in photosynthetic rate but not the three- or fourfold increase observed. Subsequently, on several occasions, photosynthetic rate was reduced below the peak value obtained around Day 35. These reductions coincided with stages in crop development when sink strength could have been limiting. The error bars on Fig. 3–5 show that the changes were statistically significant and definitely real. They are also large enough that we cannot correctly simulate crop growth if we ignore them.

My attempts to simulate these changes by allowing τ to vary as a function of the C supply/demand ratio have not been successful. Even allowing τ to vary between its extreme upper and lower limits in the course of a single day gives a model that responds in a similar manner to that observed, but responds too slowly.

ADAPTATION TO CARBON DIOXIDE CONCENTRATION

When plants are grown in two different CO_2 concentrations and their photosynthetic response to light flux density is then measured in the same CO_2 concentration, the two curves can be quite different. The curves often lie on top of each other, however, and this can lead to the conclusion that

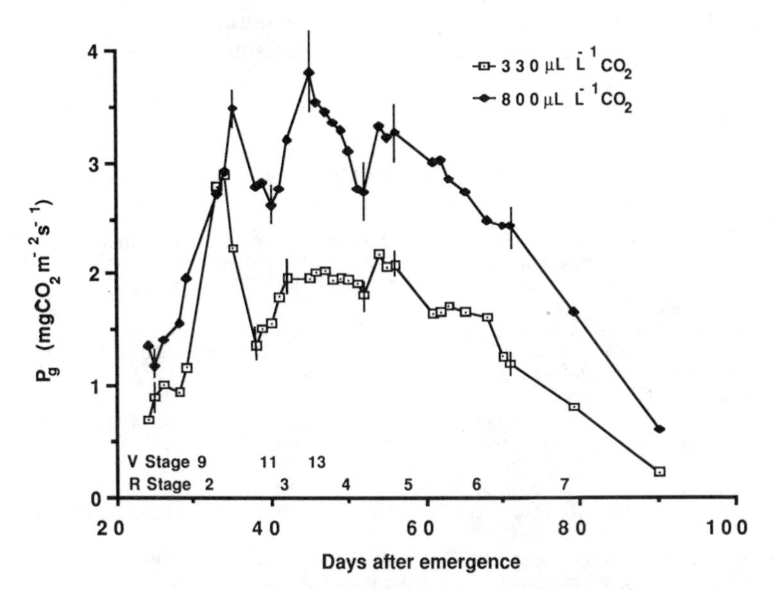

Fig. 3–5. Seasonal variation in canopy gross photosynthetic rate (P_g) at a photon flux density of 1300 μmol photon m^{-2} s^{-1} for soybean grown in 330 and 800 μL L^{-1} CO_2 concentrations. The points were estimated from curves fitted to each day's data and the 95% error bars were derived from the standard errors of the estimates.

no photosynthetic adaptation to CO_2 concentration has occurred. Closer inspection reveals that R is higher in the tissue grown in the high CO_2 concentration and that α is greater. There is also a tendency for τ to be smaller for plants grown in high CO_2. Now, we know that leaves on plants grown in high CO_2 concentrations often acquire additional mesophyll, usually an extra layer of palisade cells, and the changes observed in the photosynthetic parameters can be explained on the basis of changes in leaf morphology.

Additional mesophyll cells may increase α merely because there is more chlorophyll per unit leaf area. At high rates of photosynthesis, however, the more tortuous and longer pathway for gaseous diffusion to the mesophyll cells could decrease τ. At present, we have too little data to understand what, if any, adaptation to high CO_2 concentration is occurring, but we need to understand these phenomena if we are to predict correctly the response of crop plants to future climates with high CO_2 concentrations.

CONCLUSIONS

The simple model described in this chapter, which is empirical at the leaf level and mechanistic above that level of aggregation, is sufficient to describe the relationship between single-leaf and canopy photosynthesis and to correctly describe the response of photosynthetic rate to light flux density, CO_2 concentration, temperature, and the light regime in which the plant has developed. The attempts to deal with effects of leaf N concentration, leaf senescence, leaf water relations, and sink strength are inadequate, largely because there are too few data to reach an understanding of the mechanisms involved. Photosynthetic adaptation to CO_2 concentration has not yet been addressed in the model.

REFERENCES

Acock, B., and L.H. Allen, Jr. 1985. Crop responses to elevated carbon dioxide concentrations. p. 54–97. *In* B.R. Strain and J.D. Cure (ed.) Direct effects of increasing carbon dioxide on vegetation. U.S. Dep. of Energy DOE/ER-0238. U.S. Gov. Print. Office, Washington, DC.

Acock, B., D.A. Charles-Edwards, D.J. Fitter, D.W. Hand, L.J. Ludwig, J. W. Wilson, and A.C. Withers. 1978. The contribution of leaves from different levels within a tomato crop to canopy net photosynthesis: An experimental examination of two canopy models. J. Exp. Bot. 29:815–827.

Acock, B., V.R. Reddy, H.F. Hodges, D.N. Baker, and J.M. McKinion. 1985a. Photosynthetic response of soybean canopies to full-season carbon dioxide enrichment. Agron. J. 77:942–947.

Acock, B., V.R. Reddy, F.D. Whisler, D.N. Baker, J.M. McKinion, H.F. Hodges, and K.J. Boote. 1985b. The soybean crop simulator GLYCIM: Model documentation 1982. USDA Publ. PB85 171163/AS. USDA, Washington, DC.

Acock, B., J.H.M. Thornley, and J.W. Wilson. 1971. Photosynthesis and energy conversion. p. 43–75. *In* P.F. Wareing and J.P. Cooper (ed.) Potential crop production. Heinemann Educational Books, London.

Acock, B., J.H.M. Thornley, and J.W. Wilson. 1970. Spatial variation of light in the canopy. p. 91–102. *In* I. Setlik (ed.) Prediction and measurement of photosynthetic productivity. PUDOC, Wageningen, the Netherlands.

Björkman, O., and B. Demmig. 1987. Photon yield of O_2 evolution and chlorophyll fluorescence characteristics at 77K among vascular plants of diverse origins. Planta 170:489-504.

Boote, K.J., R.N. Gallaher, W.K. Robertson, K. Hinson, and L.C. Hammond. 1978. Effect of foliar fertilization on photosynthesis, leaf nutrition, and yield of soybeans. Agron. J. 70:787-791.

Bunce, J.A., D.T. Patterson, M.M. Peet, and R.S. Alberte. 1977. Light acclimation during and after leaf expansion in soybean. Plant Physiol. 60:255-258.

Charles-Edwards, D.A. 1978. An analysis of photosynthesis and productivity of vegetative crops in the United Kingdom. Ann. Bot. (London) 42:717-731.

Charles-Edwards, D.A., and L.J. Ludwig. 1975. The basis of expression of leaf photosynthetic activities. p. 37-44. *In* R. Marcelle (ed.) Environmental and biological control of photosynthesis. Junk, The Hague.

Clough, J.M., M.M. Peet, and P.J. Kramer. 1981. Effects of high atmospheric CO_2 and sink size on rates of photosynthesis of a soybean cultivar. Plant Physiol. 67:1007-1010.

Davidson, J.L., and J.R. Philip. 1958. Light and pasture growth. p. 181-187. *In* Climatology and microclimatology. Proc. Symp. Arid Zone Res. 11th, Canberra, ACT, Australia. UNESCO, Paris.

Ho, L.C. 1978. The regulation of carbon transport and the carbon balance of mature tomato leaves. Ann. Bot. (London) 42:155-164.

Hofstra, G., and J.D. Hesketh. 1969. Effects of temperature on the gas exchange of leaves in the light and the dark. Planta 85:228-232.

Hozumi, K., H. Kirita, and M. Nishioka. 1972. Estimation of canopy photosynthesis and its seasonal change in a warm-temperate evergreen oak forest in Minimata (Japan). Photosynthetica 6:158-168.

Iqbal, M. 1983. An introduction to solar radiation. Academic Press, Ontario, Canada.

Johnson, I.R., and J.H.M. Thornley. 1984. A model of instantaneous and daily canopy photosynthesis. J. Theor. Biol. 107:531-545.

Jurik, T.W., J.A. Weber, and D.M. Gates. 1984. Short-term effects of CO_2 on gas exchange of leaves of bigtooth aspen (*Populus grandidentata*) in the field. Plant Physiol. 75:1022-1025.

Kira, T. 1975. Primary production in forests. p. 5-40. *In* J.P. Cooper (ed.) Biosynthesis and productivity in different environments. Cambridge Univ. Press, London.

Laing, W.A., W.L. Ogren, and R.H. Hageman. 1974. Regulation of soybean net photosynthetic CO_2 fixation by the interaction of CO_2, O_2, and ribulose 1,5-diphosphate carboxylase. Plant Physiol. 54:678-685.

List, R.J. (ed.) 1963. Smithsonian meteorological tables. 6th ed. Smithsonian Inst., Washington, DC.

Ludwig, L.J., and A.C. Withers. 1978. Effect of temperature on single leaf photosynthesis and respiration in tomato. p. 51-52. *In* Rep. Glasshouse Crops Res. Inst. 1977. Littlehampton, England.

Osmond, C.B., O. Björkman, and D.J. Anderson. 1980. Physiological processes in plant ecology: Towards a synthesis with atriplex. Springer-Verlag, Berlin.

Penning de Vries, F.W.T. 1975. Use of assimilates in higher plants. p. 459-480. *In* J.P. Cooper (ed.) Photosynthesis and productivity in different environments. Cambridge Univ. Press, Cambridge, UK.

Saeki, T. 1960. Interrelationships between leaf amount, light distribution and total photosynthesis in a plant community. Bot. Mag. 73:55-63.

Saeki, T. 1963. Light relations in plant communities. p. 79-92. *In* L.T. Evans (ed.) Environmental control of plant growth. Academic Press, London.

Verhagen, A.M.W., J.H. Wilson, and E.J. Britten. 1963. Plant production in relation to foliage illumination. Ann. Bot. (London) 26:627-640.

Woodward, R.G. 1976. Photosynthesis and expansion of leaves of soybean grown in two environments. Photosynthetica 10:274-279.

4 Modeling Photosynthesis and Water-Use Efficiency of Canopies as Affected by Leaf and Canopy Traits

Vincent P. Gutschick
New Mexico State University
Las Cruces, NM

The physiological-micrometeorological modeling I describe here guides our experimental program in physiological crop breeding for water-use efficiency (V.P. Gutschick, C.G. Currier, and G.L. Cunningham, U.S. Geological Survey Grant 14-08-0001-G1641.) and our allied ecological research. I will allude to several related modeling efforts but will concentrate on modeling water-use efficiency (WUE) and yield in alfalfa (*Medicago sativa* L.). This particular model, which I call FORWUEY (FORage WUE and Yield), is designed to be sufficiently mechanistic, inclusive in subprocesses, and accurate to predict how WUE and yield depend on complexes of traits. The principal traits are C_i (leaf internal CO_2 concentration) and plant-average SLM (specific leaf mass). I also address how WUE and yield depend on leaf zenith angle and on stomatal sensitivity to vapor-pressure deficit (VPD), as well as on environmental conditions. While the model is developed specifically for alfalfa, its form and probably its (promising) conclusions should apply widely to other forage crops. In fact, the method of improving WUE by breeding for lower C_i and higher SLM should apply to most reproductive-yield crops as well, because C_i and SLM have no apparent linkage to vegetative-reproductive allocation patterns.

Such predictive modeling for crop design is quite different from, and complementary to, the preponderance of modeling for crop management (e.g., Wisiol & Hesketh, 1987). The crop-design model here states explicitly the quantity to be optimized, a mixture of WUE and yield. Key subprocess models are directly related to heritable or presumably heritable traits. In particular, the stomatal conductance is modeled as being controlled by mesophyll conductance (in turn, determined by irradiance and SLM) and by mechanisms of unknown detail that maintain a constant ratio of C_i to external CO_2 concentration (Wong et al., 1985a,b,c; Küppers et al., 1986). The model's resolution of a number of processes, particularly in the soil, is restricted.

 Modeling Crop Photosynthesis—from Biochemistry to Canopy. CSSA Special Publication no. 19.

The model has been used in several ways. Foremost, I have used it to estimate potential performance gains, e.g., a 25% increase in WUE with a modest 10% penalty in yield. In this respect, its use is analagous to the past use of models to predict yield gains from greater leaf erectness (de Wit, 1965; Duncan et al., 1967; Loomis & Williams, 1969), building on earlier suggestions of Boysen Jensen (1932) and of Monsi and Saeki (1953). This prediction led to breeding experiments that met with some success, as reviewed by Trenbath and Angus (1975). The model I present also predicts the exact form of the nonlinear dependence of WUE and yield on C_i and SLM, apparent in Fig. 4-1 through 4-4. As a corollary, the model predicts the relative sensitivity of experimental results (WUE and yield directly, or the potential to alter them by selecting variant C_i and SLM) to measurement errors and to environmental factors; thus, it helps design experimental protocols.

WHY ATTEMPT PHYSIOLOGICAL BREEDING FOR WATER-USE EFFICIENCY?

We propose that WUE may be selected indirectly by selecting for two physiological traits, C_i and SLM. Compared with direct selection, such a program may demand extra effort in field measurements and definitely demands a firm knowledge of which traits control WUE and yield, and how they do. However, physiological breeding bears several advantages (cf. Falconer, 1960; Wallace et al., 1972):

1. Fewer inadvertent changes are likely to be made in other aspects of performance. For example, in selecting for WUE defined as shoot dry matter produced per unit water used, direct selection at high water availability might inadvertently select some genotypes with reduced root/shoot ratio that aids WUE but penalizes drought tolerance; indirect selection for C_i and SLM is unlikely to co-select low root/shoot ratio.
2. Less overall genetic variation is carried. Gains in performance might thus be achieved faster and might be more stable.
3. Selection trials are less biased by environment, in general.

The expectation that WUE may be improved by indirect or direct selection has three bases. First, the WUE of agricultural crops has rarely been selected, in contrast to yield, which has been heavily selected and may be reaching a plateau. Even the selection of dryland yield may have improved drought tolerance more than WUE. Thus, even with the realization that WUE has some rather strict upper limits (Jones, 1983; Sinclair et al., 1984), there may be room for significant improvement—perhaps the 25% my model predicts. (Actual gain is a function of the yield penalty one is willing to accept.) Second, alfalfa itself has been observed (Wilson et al., 1983; Currier et al., 1987) to vary significantly in productivity, a mixed function of WUE and drought tolerance, under limited water availability. Third, the selection of lower C_i, which is a key part of our proposed breeding strategy, appears to be effective in improving WUE in other crops such as peanut (*Arachis hypogaea* L.; Hubick et al., 1986), tomato (*Lycopersicon esculentum* L.; Martin et al., 1989), and range grasses (Johnson et al., 1988). In fact, our proposal to select conjointly for higher SLM can be viewed as a way to amplify gains from selecting low C_i or to reduce the associated yield penalties (see below).

Model Inputs: The Physiological, Morphological, and Phenological Bases of Water-Use Efficiency

The primacy of the environment in determining WUE is uncontested (Jones, 1983). Among plant factors, the primacy of photosynthetic pathway in determining WUE is likewise uncontested. Nevertheless, there are other traits in which variations contribute to selectable variations in WUE. These traits are variously physiological (especially variations in C_i within the C_3 pathway), morphological (e.g., leaf size and shape; Ferguson, 1974; Hiebsch et al., 1976), and phenological (e.g., earliness, in a Mediterranean environment; Fischer and Turner, 1978).

The initial development of my model for WUE gains, and more generally for optimizing the tradeoff of WUE and yield according to economic conditions, may be outlined as follows:

1. Lower C_i leads to increased WUE, measured as short-term gas exchange (the ratio of CO_2 gained to water vapor transpired; this will be abbreviated as WUE_{gx}). This follows from the biophysical relation of photosynthetic and transpiration rates. Both rates are proportional to gas-concentration differences across a similar total resistance (see Jones, 1983):

$$WUE_{gx} \approx \frac{(C_a - C_i)}{[e^{sat}(T_l) - e_a]}\frac{1}{1.6} \approx \frac{C_a(1 - C_i/C_a)}{VPD}\frac{1}{1.6}. \quad [1]$$

This is a transpirational WUE rather than a total WUE including soil evaporation. Here, e^{sat} and e_a are water-vapor concentrations, the saturated value in the leaf interior and the ambient-air value, respectively, C_a is the ambient CO_2 concentration, and T_l is the temperature at the leaf surface. Given that the ratio C_i/C_a, which I will call τ, is held quite stable by uncertain physiological mechanisms as the environment varies (Wong et al., 1985a,b,c; Küppers et al., 1986), selection for lower τ will improve WUE. The gain is compromised because a lower τ, all else being equal (especially SLM), is achieved by lower stomatal conductance, g_s. This entails lower transpirational cooling and higher VPD. The selection process is complicated because $C_i(\tau)$ responds to VPD itself (Lösch & Tenhunen, 1981; Mansfield & Davies, 1985), to leaf water potential (Forseth & Ehleringer, 1983; Henson et al., 1989), and to leaf age (Vos & Oyarzún, 1987).

2. Decreased C_i also lowers the light-saturated rate of photosynthesis per unit leaf area, which I denote as P_{la}^{max}. At τ values near 0.7 that typify C_3 plants, a given fractional change in τ generates about the same fractional change in P_{la}^{max}. This is seen in experimental studies (e.g., Badger et al.,

1984) and is implicit in the enzyme-kinetic analyses of C_3 photosynthesis (Farquhar et al., 1980).

3. Higher SLM may raise P_{la}^{max} nearly proportionally (Pearce et al., 1969; Kallis & Tooming, 1974; Nelson, 1988). However, the gain in photosynthesis averaged across the entire leaf area of the canopy is smaller. Leaves in different positions in the canopy experience a variety of irradiances, many below light saturation, and thus the gain in average P_{la} is less than proportional to SLM. More importantly, higher SLM at a given plant size (mass, M_p) entails a reduction in leaf area. The net effect on canopy photosynthesis (P_{can}) is negative in early growth and positive at high leaf-area index, LAI. The trade-off in net yield is complex and its prediction requires a whole-season, whole-canopy model such as I use here. The trade-off also points out that measures of photosynthetic performance must be chosen carefully in experimental studies (Gutschick, 1987a,b).

4. Higher SLM should also modestly increase WUE directly. All else being equal (constant τ, in particular), higher SLM leads to higher rates of photosynthesis and transpiration. The leaves are cooled more and have a lower temperature, hence a lower VPD.

The couplings between C_i and leaf temperature demand a more inclusive model that includes energy balance. A full-canopy model with accurate description of light penetration and of gas transport is demanded by a number of factors, including the variations of irradiance, temperature, water-vapor concentration, and CO_2 concentration within the canopy, especially by depth, and the couplings of these four variables to transpiration, and hence to SLM and τ.

THE FORWUEY MODEL

The model simulates the vegetative growth of alfalfa in standard repeated cutting cycles. It predicts harvest-basis WUE (shoot dry matter produced, divided by cumulative transpiration) and yield, both as functions of C_i and SLM in an otherwise fixed genetic background, in a given, stationary, stochastic mix of environmental conditions. Yield is predicted indirectly, from t^*, the time required for a stand to regrow from shoot biomass density M_0 immediately after harvest to a final biomass density, M_f. In a long run of seasons of average duration t_{season}, the yield, Y, of a stand is estimated as

$$Y = \text{(average number of cuts)(mass harvested per cut)} \quad [2]$$

$$\approx \frac{t_{season}}{t^*}(M_f - M_0).$$

Thus, yield is inversely proportional to regrowth time. The values of M_0 and M_f are presumed not to vary between genotypes differing in C_i and SLM. We have no information yet on whether this assumption holds in field growth. This assumption probably gives a more conservative estimate of yield penalties (that is, overestimates them) than the other limiting assumption that regrowth time is invariant.

The crop canopy is assumed to be laterally homogeneous and the leaves are assumed to be distributed randomly in space (the Poisson distribution of gap probabilities: Ross, 1981).

The variable physiological traits, subject to selection, are taken to be SLM and τ. The fixed aspects of the plant physiology and morphology are specified by 11 parameters. Parameter 1, the leaf angle, θ_l, is assumed fixed at 20° from horizontal. In reality, upper-canopy alfalfa leaves track the sun diaheliotropically (Scott & Wells, 1969; Travis & Reed, 1983; Reed & Travis, 1987). I have run a modified version of the model with complete diaheliotropism to estimate effects of tracking on WUE and yield, as reported below.

Four more parameters describe the leaf area at a given growth stage (Parameters 2, 3, and 4). Three parameters, over and above SLM, describe the leaf area per plant, L_p, as a function of the leaf mass per plant, m_l, which in turn is a function of shoot mass per plant, m_s, and thus of total mass per plant, m_p:

$$L_p = \frac{m_l}{\text{SLM}} \quad [3a]$$

$$\ln m_l \approx \left\{\frac{5}{6} - \frac{1}{6}\tanh[a(\ln m_s - b)]\right\} \ln m_s + y_0. \quad [3b]$$

The shoot mass is given in terms of the root/shoot ratio r as $m_s = m_p r/(1 + r)$. At low m_s, the ratio m_l/m_s is a constant, e^{y_0}. Asymptotically at high m_s, m_l scales up as $m_s^{2/3}$, with a transition region set by values of a and b. The values $a = 0.182$, $b = 6.641$, and $y_o = -0.511$ roughly fit our field data on alfalfa. Parameter 5: A fixed ratio of root/shoot growth increments is used, $r = 0.2$ (clearly a simplification, in light of data of Baysdorfer & Bassham, 1985; Feltner & Massengale, 1965; Versteeg et al., 1982).

Four parameters (p_{CS}, Q_{00}, ϕ_c, and θ_{PS}) describe how leaf photosynthetic rate and stomatal conductance depend on C_i, SLM, and irradiance. The dependence of P_{la} on irradiance, I_l, is modeled after Johnson and Thornley (1984):

$$P_{la} = \frac{1}{2\theta_{PS}}\left\{\frac{I_l}{Q_0} + P_{la}^{max} - \left[\left(\frac{I_l}{Q_0} + P_{la}^{max}\right)^2 - \frac{4\theta_{PS} I_l P_{la}^{max}}{Q_0}\right]^{1/2}\right\} \quad [4]$$

Here, θ_{PS} is a rectangularity parameter, with limits 0 (the rectangular hyperbola) and 1 (a Blackman curve). A value of 0.9 closely fits our alfalfa gas-exchange data. Following Farquhar et al. (1980), the initial quantum yield, Q_0, is taken as a function of the oxygenation/carboxylation ratio, ϕ;

$$Q_0 = Q_{00}\,\frac{(1 - 0.5\phi)}{(1 + \phi)}. \quad [5]$$

In turn, ϕ is a function of C_i;

$$\phi = \phi_C / C_i. \qquad [6]$$

At an assumed constant O_2 concentration of 4.2 mol m^{-3}, ϕ_C equals 2.52 mmol m^{-3}, equivalent to 61.8 μmol mol^{-1} of CO_2 at 25° C. The value of CO_2-saturated quantum yield, Q_{00}, is taken as 0.0732 mol CO_2 mol^{-1} photon, giving $Q_0 = 0.05$ in the same units when $C_i = 0.0092$ mol m^{-3} (230 μmol mol^{-1}, typical of C_3 plants; Bell, 1982). The value of P_{la}^{max} is taken as

$$P_{la}^{max} = p_{CS} C_i \mathrm{SLM}, \qquad [7]$$

with $p_{CS} = 1.07$ m^3 g^{-1} s^{-1} to fit field data of Pearce et al. (1969) and Kallis and Tooming (1974).

Two further parameters describe photoassimilate partitioning in growth and maintenance. Parameter 10: the biosynthetic conversion efficiency, β, is taken as 0.67 g dry matter g^{-1} glucose, following Penning de Vries et al. (1974). Parameter 11: the specific maintenance coefficient, a_{maint}, is taken as 0.03 g glucose g^{-1} dry matter d^{-1} (Penning de Vries, 1975), equally for all active tissue (shoot and fraction of root).

Plant density is taken as 75 m^{-2} for a mature stand. For simplicity, temperature effects on processes other than transpiration are neglected; parametrizations are available for later extensions of the model, although they are not fully satisfactory. Stomatal conductance is assumed insensitive to VPD in the basic model, and sensitive as a ramp function in an extended model, FORWUEY.VPD. In addition to all the above, the model incorporates 11 physical and micrometeorological constants, a mean leaf absorbance of photosynthetically active radiation (0.85), and 10 error-control parameters for numerical solution techniques.

The environment of growth is described as a stochastic mix of a number, N_e, of discrete conditions; the ith occurs with a probability w_i. I use four conditions, three to sample the diurnal cycle on a clear day and one to represent overcast conditions. Each condition i is described by: (i) I_{00} = the direct solar-beam irradiance normal to its direction of propagation; (ii) θ_s = the solar elevation angle; (iii) D_{00} = the diffuse sky irradiance on a horizontal surface; (iv, v, and vi) T_a^0, e_a^0, and C_a^0, which are, respectively, the air temperature, water-vapor concentration, and CO_2 concentration at the top of the canopy (or atop the canopy boundary layer, if one exists); (vii) u^0 = wind velocity at the top of the canopy; (ix) T_{sky}^{eff} = effective radiant temperature of the sky in the thermal infrared (Jones, 1983). Again for simplicity in seeing effects, soil hydrology is simplified to no exchange of CO_2, water vapor, or heat between soil and air.

The model has been improved in its descriptions of the temperature dependence of photosynthesis, the humidity sensitivity of stomatal conductance, and soil evaporation.

PROCESS MODELING

At each of 21 discrete shoot biomass densities, M, and for each environmental condition i, the model calculates instantaneous rates of $P_{can}(M,i)$ and canopy transpiration, $E_{can}(M,i)$. The gas exchange is assumed to be in steady state (no transients) and an absence of water stress is assumed (this latter effect is generalizable). Then, $P_{can}(M,i)$ and $E_{can}(m,i)$ are averaged across i to yield $\overline{P}_{can}(M)$ and $\overline{E}_{can}(M)$. Such simple averaging disallows modeling the effect of transpirational water loss on plant water potential that is cumulative through the day, and the reciprocal effect of water potential on photosynthesis and transpiration. Also, the growth trajectory for an averaged environment is not quite the same as the average growth trajectory for separate environments (Jones, 1981). Both limitations can be remedied by straightforward extensions of the model that retain its computer-time-saving method of indirectly integrating the growth equation in time, below. Finally, the growth equation (top line, below) is integrated by a useful transformation (second line, below):

$$\frac{dM}{dt} = \beta[\overline{P}_{can}(M) - a_{maint}M] \qquad [8]$$

$$t^* = \int_{M_0}^{M_f} dt = \int \frac{dM}{\beta[\overline{P}_{can}(M) - a_{maint}M]} \equiv \int dMQ(M).$$

The integrand $Q(M)$ at the 21 different M values is fitted to a Chebyshev polynomial and integrated trivially, saving much time compared with stepwise integration of M in small time steps. A related transformation similarly simplifies the calculation of season-average WUE.

In order to calculate $P_{can}(M,i)$, the model first calculates the mass per plant as the shoot biomass density divided by the number density of plants per ground area. Then the allometric relations yield L_p (Eq. [3]), hence the canopy LAI. Next, leaf gas exchange is calculated at each of 21 discrete depths (values of cumulative LAI, L). A sophisticated model that predicts light penetration and projection on inclined leaf surfaces calculates the probability distribution, $P(I_l)$, of irradiance I_l on leaf area. The light-distribution model accounts for contributions from direct and diffuse sunlight and from light scattered from other leaves and soil. For calculations of energy balance later, an analogous model computes the distribution of near-infrared radiation on leaves.

For a given leaf area receiving a given irradiance I_l, the photosynthetic rate is calculated from Eq. [4] and the auxiliary relations in Eq. [5] through [7]. Using these equations requires that C_i be known. The premise of the model is that

$$C_i = \tau C_i \qquad [9]$$

under any CO_2 concentration and any but the lowest irradiance. Here, C_i is the CO_2 concentration at the leaf surface, below the leaf boundary layer:

$$C_i = C_a - r_{b,leaf} P_{la}. \quad [10]$$

The composite equation,

$$C_i = \tau[C_a - r_{b,leaf} P_{la}(C_i)], \quad [11]$$

must be solved iteratively numerically for C_i. A second complication is that C_a is not known a priori because it depends on the photosynthetic activity elsewhere in the canopy and on canopy transport properties. The model, therefore, calculates leaf performance for a reference CO_2 distribution, $C_a(L)$, and also calculates the derivative of performance with respect to C_a. The total solution for $C_a(L)$ and, therefore, for P_{can} is iteratively improved.

At the same time that leaf photosynthetic performance is being calculated, so, too, is leaf energy balance and, hence, leaf temperature. Steady-state temperature is calculated, using contributions from shortwave radiation, incoming thermal infrared (TIR—from sky, leaves, soil), outgoing TIR, transpiration, and convective-conductive heat transfer. For transpirational cooling rate, the stomatal conductance for water vapor is scaled from the conductance g_{sC} for CO_2. In turn, g_{sC} is set by the photosynthetic rate,

$$g_{sC} = \frac{P_{la}}{C_l - C_i}. \quad [12]$$

The nonlinear energy-balance equation is solved iteratively, because leaf temperature occurs in both exponential and linear functions. Energy balance also depends on leaf photosynthetic activity that determines conductance. Now, P_{la} (and thus energy balance) depends on the depth-dependent concentration $[C_a(L)]$, and in general on the air temperature $[T_a(L)]$ and water-vapor content $[e_a(L)]$ as well. These three functions depend on P_{la}, etc., at all depths (see below), making a closed loop of calculations that must be solved iteratively.

Given the statistically averaged photosynthetic and transpiration rates at each of the 21 discrete canopy depths, the model then solves the transport equations for CO_2, water vapor, and heat, using simple K-theory (Cowan, 1968), augmented by accounting for a canopy boundary layer (Jarvis & McNaughton, 1986). The solutions for $T_a(L)$, $C_a(L)$, and $e_a(L)$ are each represented by Chebyshev series and solved quickly; they are improved iteratively. The spatial derivatives of C_a and e_a yield the canopy photosynthetic and transpirational rates.

The model is coded in strict Fortran 77 and is completely portable (machine independent). I run it on a Sun 3/75M-4 workstation (32-bit CPU, 16 MHz, about 2 MIPS) with a Motorola 68881 math co-processor. A simulation of one growth cycle with four different growth conditions requires about 12 min for each choice of C_i and SLM.

MODEL RESULTS

The most compact display of predicted WUE and yield is as contour plots (Fig. 4–1), with $\tau = C_i/C_l$ s one axis and SLM as the other axis. In predicting useful breeding efforts, I use the plots four ways:

1. Consider the goal of preserving yield unchanged and finding the combination of C_i and SLM that gives the highest WUE. First, I locate the current cultivar, say, Mesilla, according to its C_i and SLM (the asterisk in Fig. 4–1). The short, dotted line curving to the right follows an intermediate contour of constant regrowth time, hence of putatively constant yield. I terminate this line at the point where it reaches the maximal WUE. The projected gain is a very modest 4%.

2. Consider the goal of improving WUE to some chosen value and doing so with the least penalty in yield. For example, I have chosen a 25% gain in WUE and have drawn (not shown) that WUE contour. Then I search along the WUE contour for the place where it is tangent to a contour of t^*, which will be the shortest t^* possible. The shift in t^* is seen to be +12%, corresponding to a drop in average yield of $(1 - 1/1.12) \times 100\% = 11\%$ (and a 33% reduction in total water use as transpiration). Achieving this requires a drop in τ from 0.85 to 0.75 (possible, as our experimental results

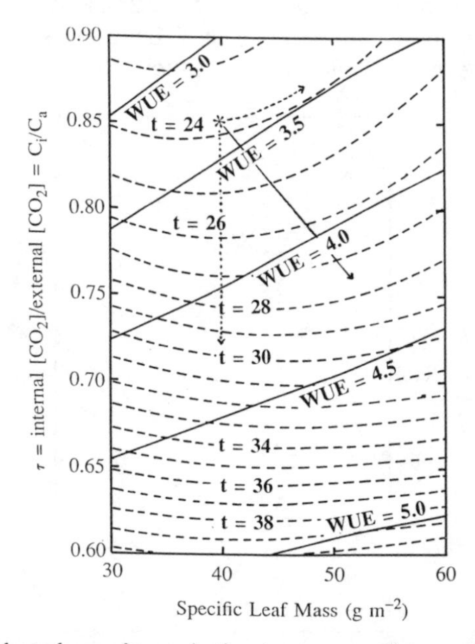

Fig. 4–1. Predicted dependence of transpirational water-use efficiency (WUE, in g dry matter kg^{-1} water, solid lines) and regrowth time (in d, dashed lines) on C_i and specific leaf mass, for alfalfa regrowing from 50 to 500 g m^{-2} in shoot biomass density. A mix of four environmental conditions was used to simulate late-spring conditions in southern New Mexico. The asterisk indicates C_i and specific leaf mass of Mesilla cultivar. See text for the significance of lines from the asterisk indicating various breeding strategies.

show) and an increase in SLM from 40 to 50 g m^{-2}, also possible with the Mesilla gene pool. (Similar gains and requirements apply to the new cultivar Wilson, on which we also have done both modeling and experiments.)

3. Consider, finally, an arbitrary mixed goal of gaining x% in WUE and changing yield y%, perhaps indicated as optimal in economics by linear programming. One can readily devise a search on the contour plot.

4. Compare the projected gains in Method 2 above with the gains obtainable if one only breeds for lower C_i. The downward, vertical dotted line terminates at a combination of C_i and SLM that also gives 25% higher WUE; the associated penalty in average yield is estimated as 16%. Adding SLM to the selection scheme, as in Method 2, cuts the yield penalty from 16% to 11%, with very little increase in breeding effort and cost.

I have run the model for a number of other environmental conditions and modifications of plant physiology and morphology:

1. When temperatures are increased while relative humidity remains constant, WUE drops overall, as expected. Projected combinations of relative increases in WUE and decreases in yield are not strongly altered.

2. Introduction of a significant canopy boundary-layer resistance (Fig. 4–2) raises all WUEs, decreases all yields (relatively less than WUE is increased), and compresses WUE differences among all C_i–SLM combinations.

3. Figure 4–3 shows the effects of introducing a realistic ramp response of stomatal conductance to VPD. The WUE is raised for all C_i–SLM combinations, as expected, but not very much. Yield is cut proportionally

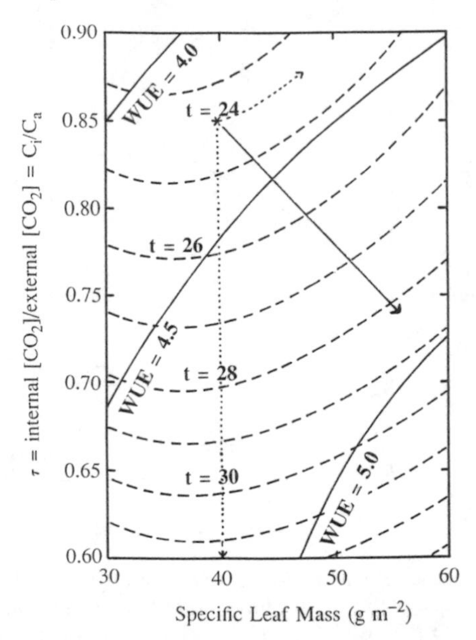

Fig. 4–2. Predicted dependence of transpirational water-use efficiency and regrowth time on C_i and specific leaf mass when canopy boundary-layer resistance is 25 s m^{-1}. All other environmental conditions and plant parameters are as in Fig. 4–1.

more, especially at low τ and low SLM, where transpirational cooling is poorest; the VPD tends to a runaway behavior in these hypothetical plants with hot leaves.

4. Accounting for SLM grading from high values at the top of the canopy to low values deep in the canopy gives results like those in Fig. 4–4. In this more realistic simulation, yield is greater than in the simpler simulations without SLM gradation—by 10% at low C_i and SLM, negligibly at high SLM. This is consistent with the results of Gutschick and Wiegel (1988). The WUE is decreased about 5%, for complicated reasons. The potential advantage of C_i–SLM selection over C_i-only selection may be substantially cut.

5. Hypothetical breeding for erect leaves (70°) improves yield of low C_i–low SLM genotypes (most readily light saturated) and penalizes yield of high C_i–high SLM combinations. WUE is increased, modestly for the latter genotypes that effectively cool themselves transpirationally, and significantly for the former genotypes that do not cool themselves well.

6. Perfect diaheliotropic solar tracking by leaves decreases WUE slightly, by 6 to 7% for any C_i and SLM. Yield is not improved for C_i and SLM values typical of modern alfalfa cultivars. Several other cases have been run; the information in all the runs has yet to be capitalized on in our experimental work.

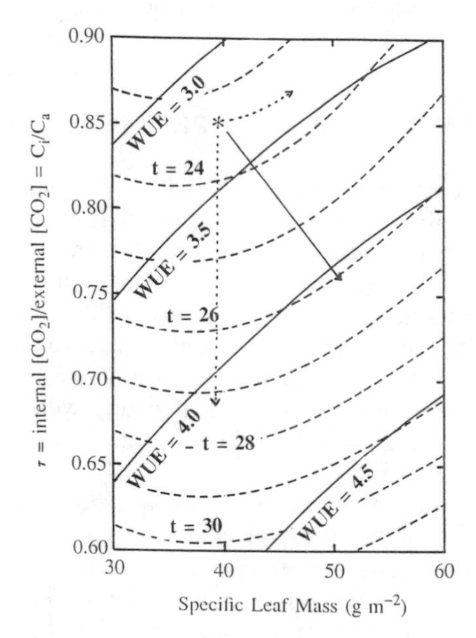

Fig. 4–3. Predicted dependence of transpirational water-use efficiency and regrowth time on C_i and specific leaf mass when stomatal conductance decreases linearly with vapor-pressure deficit (VPD) exceeding 0.4 kPa, reaching a limiting value of 0.1 times the value that preserves C_i when VPD exceeds 1.6 kPa. All other environmental conditions and plant parameters are as in Fig. 4–1.

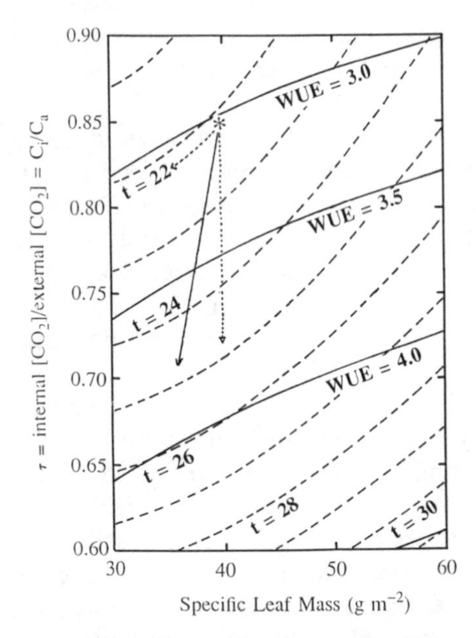

Fig. 4–4. Predicted dependence on transpirational water-use efficiency and regrowth time on C_i and canopy-average specific leaf mass, when specific leaf mass (SLM) decreases linearly from the top of the canopy by 15 g m^{-2} per unit leaf area index, until SLM reaches a minimum of 25 g m^{-2}. All other environmental conditions and plant parameters are as in Fig. 4–1.

A BRIEF COMPARISON WITH EXPERIMENTAL RESULTS

Our experimental program (Gutschick and Cunningham, 1989) has five objectives: (i) to measure the variations in C_i and SLM among individual plants (genotypes) in two cultivars, Mesilla and Wilson; (ii) to measure the variations in WUE and yield in these same plants; (iii) to correlate the variations in WUE and yield with the variations in C_i and SLM in order to test the accuracy of the model's predictions and in order to estimate potential gains in WUE; (iv) to determine heritabilities of C_i and SLM, particlarly in different growth environments such as field plants will experience; and (v) to extend all these inquiries to field environments, where processes unaccounted in growth chambers may enter.

We grow 40 individual plants of each cultivar in dense swards in growth chambers that we have modified (Pushnik et al., 1988) to achieve field-realistic high irradiances without adverse effects of high TIR loading of leaves. Temperature, humidity, and photoperiod are controlled. Water use of individual plants is measured by mass balance and yield is measured by harvest. At two periods during each growth cycle, we measure C_i by gas exchange and also by the $^{13}C/^{12}C$ ratio in leaf tissue that indicates longer term average C_i (Evans et al., 1986). The SLM is measured destructively.

In our first growth cycles, the observed ranges in C_i and SLM were large enough both to test predicted correlations with WUE and yield and to use in breeding. In the direct measurements of gas exchange, the range in C_i expressed as $1 - C_i/C_a$ (most relevant for WUE) was 0.07 to 0.17 in Mesilla; in Wilson, it was 0.15 to 0.27. The range in SLM was 17.2 to 26.5 g m^{-2} in Mesilla and from 17.3 to 33.9 g m^{-2} in Wilson; this is below field averages, for various reasons, but the ranges are usefully large. Interestingly, C_i and SLM appear to be negatively correlated (with $P \approx 0.04$), as one could expect if increased SLM increases the leaf biochemical capacity or mesophyll conductance without a significant accompanying alteration in stomatal conductance. This tie is beneficial in the anticipated breeding effort, in which higher SLM and lower C_i are sought. The ranges in WUE in each cultivar are approximately twofold, from 2.23 to 3.63 mg dry matter g^{-1} water in Mesilla, and from 1.46 to 2.92 mg dry matter g^{-1} water in Wilson growing at a lower humidity. The relative ranges are similar to the 1.7-fold range seen in various *Arachis* species by Hubick et al. (1986) in a study of C_i–WUE relations for breeding. Yield of individual plants varies fivefold, far more than the range predicted to arise from C_i and SLM variations. Undoubtedly, differences in inherent vigor and in C_i and SLM-dependent growth rates are amplified exponentially by competitive growth of the genotypes. The WUE correlates negatively with C_i at fairly low levels of statistical significance:

$$\text{Mesilla: } \mathrm{WUE_{shoot}} = -7.23\tau + 9.14 \text{ (mg dry matter g}^{-1}\text{ water)} \qquad [13]$$

$$r = -0.532,\ n = 14,\ P \approx 0.04$$

$$\text{Wilson: } \mathrm{WUE_{shoot}} = 0.214\delta^{13}\mathrm{C} + 8.79 \text{ (mg dry matter g}^{-1}\text{ water)}$$

$$r = 0.435,\ n = 16,\ P \approx 0.10$$

The WUE correlated positively with SLM, again with modest significance:

$$\text{Mesilla: } \mathrm{WUE_{shoot}} = 0.0633\ \mathrm{SLM} + 1.441 \text{ (mg dry matter g}^{-1}\text{ water)} \qquad [14]$$

$$r = 0.482,\ n = 17,\ P \approx 0.04$$

Yield correlates significantly with C_i only in Wilson, and there in the direction opposite from predictions that assume SLM and C_i are uncorrelated:

$$\text{Yield} = 1.996\delta^{13}\mathrm{C} + 65.22 \qquad [15]$$

$$r = 0.776,\ n = 16,\ P < 0.001$$

Here I have used, as a measure of C_i, the C-isotope ratio, $\delta^{13}C$, relative to a mineral standard. The isotopic ratio is a better measure of time-averaged performance in the leaf than is direct gas exchange. A partial explanation of the unexpected negative correlation of yield with C_i (positive correlation with $\delta^{13}C$) is, of course, that low C_i correlates with high SLM to offset yield penalties. A fuller explanation requires that the shape of the contours of regrowth time in Fig. 4–1 be modified; several clear leads are being pursued.

CONCLUSIONS, AND LESSONS OF AND FOR MODELING

The combination of modeling and experiment encourages us, short of having completed field experiments, that usefully large gains in WUE are achievable with indirect genetic selection for C_i and SLM. A corequisite for breeding is that SLM and C_i be heritable. Heritability of SLM in alfalfa has been demonstrated by Song and Walton (1975). Heritability of C_i has been shown in tomato (Martin et al., 1989) and in the broad sense in *Arachis* (Hubick et al., 1986) and in range grasses (Johnson et al., 1989). It is tempting to speculate that breeding for lower C_i may be simplified and speeded up by using restriction-fragment mapping (Tanksley and Orton, 1983); Martin et al. (1989) have found in tomato that lower C_i correlated with a relatively few restriction fragments, possibly related to relatively few genes (for either stomatal or mesophyll contributions to control of C_i, and for SLM itself).

Simultaneous modeling of WUE and yield appears to be valuable. Modeling and experiments may reciprocally improve each other. Models in particular may aid the identification of physiological traits suited to particular environments, the effects of which are more clearly and easily simulated than untangled from field results. Models also can be suited to delineate how many modest gains (as from lower C_i, higher SLM, more nearly optimal profiles of SLM with depth) may be aggregated into significant gains in WUE or yield or other performance criteria.

My model incorporates several useful techniques that can be put into many other crop models to improve their speed, e.g., the Chebyshev polynomial techniques of summing photosynthesis and transpiration throughout a canopy and of integrating quickly in time.

A number of interesting challenges lie along the path to improved modeling of WUE and yield. There remain problems of accounting for fluctuating irradiance in canopies and of incorporating soil–plant water relations. Describing the effects of the age structure of leaves in a whole canopy will also be challenging. Going beyond the immediate question of WUE and yield in a specified population, future model extensions should address two major questions. First, how are WUE and drought tolerance (DT) related to each other physiologically and ecologically? Some physiological traits contribute to WUE and DT oppositely, e.g., a high root/shoot ratio should improve DT but, by decreasing shoot allocation, should depress shoot-basis WUE. Ecologically, some interspecies competitions appear to be stabilized if there is high WUE–low DT in one species and low WUE–high DT in the other (Barnes, 1986; Thomas, 1986; cf. Derera et al., 1969). If such dichotomy in WUE–DT combinations occur in alfalfa, how will they affect differential survival of individuals in new populations bred for higher WUE? Specifically, will highest WUE individuals be pruned away so that WUE gains decline over the years? Second, if breeding for low C_i and high SLM also proves useful in crops other than alfalfa and other legumes, there is the concern that N-use efficiency is negatively correlated with water-use efficiency (field data of Field et al., 1983; physiological arguments readily constructed). Improving WUE might demand more fertilizer use—although our proposed strategy of altering SLM amounts to redistributing a fixed mass of C or N, not adding N.

Finally, the model is potentially useful in other inquiries. Because it incorporates realistic mechanistic responses to external CO_2 levels, it may be applied in research to quantify effects of global CO_2 increases, which are expected to improve yield in most plants (Cure & Acock, 1986), improve WUE, and alter hydrological balances variously in different regions of the world.

REFERENCES

Badger, M.R., T.D. Sharkey, and S. von Caemmerer. 1984. The relationship between steady-state gas exchange of bean leaves and the levels of carbon-reduction-cycle intermediates. Planta 160:305–313.

Barnes, F. 1986. Carbon gain and water relations in pinyon–juniper habitat types. Ph.D. diss. New Mexico State Univ., Las Cruces (Diss. Abstr. 86-26082).

Baysdorfer, C., and J.A. Bassham. 1985. Photosynthate supply and utilization in alfalfa. A developmental shift from a source to a sink limitation of photosynthesis. Plant Physiol. 77:313–317.

Bell, C.J. 1982. A model of stomatal control. Photosynthetica 16:486–4495.

Boysen Jensen, P. 1932. The dry matter production of plants. G. Fischer Verlag, Jena, Germany.

Cowan, I.R. 1968. Mass, heat and momentum exchange between stands of plants and their atmospheric environment. Q.J.R. Meteorol. Soc. 94:523–544.

Cure, J.D., and B. Acock. 1986. Crop responses to carbon dioxide doubling: a literature survey. Agric. For. Meteorol. 38:127–145.

Currier, C.G., B.A. Melton, and M.L. Wilson. 1987. Evaluation of the potential to improve alfalfa for production under less than optimum moisture conditions. Tech. Completion Rep. 1423649. New Mexico Water Resourc. Res. Inst., Las Cruces.

Derera, N.F., D.R. Marshall, and L.N. Balaam. 1969. Genetic variability in root development in relation to drought tolerance in spring wheats. Exp. Agric. 5:327–337.

de Wit, C.T. 1965. Photosynthesis of leaf canopies. Agric. Res. Rep. 663. PUDOC, Wageningen, the Netherlands.

Duncan, W.G., R.S. Loomis, W.A. Williams, and R. Hanau. 1967. A model for simulating photosynthesis in plant commnities. Hilgardia 38:181–205.

Evans, J.R., T.D. Sharkey, J.A. Berry, and G.D. Farquhar. 1986. Carbon isotope discrimination measured concurrently with gas exchange to investigate CO_2 diffusion in leaves of higher plants. Aust. J. Plant Physiol. 13:281–292.

Falconer, D.S. 1960. Introduction to quantitative genetics. Ronald Press, New York.

Farquhar, G.D., S. von Caemmerer, and J.A. Berry. 1980. A biochemical model of photosynthetic CO_2 assimilation in leaves of C_3 species. Planta 149:78–90.

Feltner, K.C., and M.A. Massengale. 1965. Influence of temperature and harvest management on growth, level of carbohydrates in the roots, and survival of alfalfa (*Medicago sativa* L.). Crop Sci. 5:585–588.

Ferguson, H. 1974. Use of variety isogenes in plant water-use efficiency. Agric. Meteorol. 14:25–29.

Field, C., J. Merino, and H.A. Mooney. 1983. Compromises between water-use efficiency and nitrogen-use efficiency in five species of California evergreens. Oecologia 60:384–389.

Fischer, R.A., and N.C. Turner. 1978. Plant productivity in the arid and semi-arid zones. Annu. Rev. Plant Physiol. 29:277–317.

Forseth, I.N., and J.R. Ehleringer. 1983. Ecophysiology of two solar tracking desert winter annuals. III. Gas exchange responses to light, CO_2 and VPD in relation to long-term drought. Oecologia 57:344–351.

Gutschick, V.P. 1987a. A functional biology of crop plants. Timber Press, Beaverton, OR.

Gutschick, V.P. 1987b. Quantifying limits to photosynthesis. p. 67–87. *In* K. Wisiol and J.D. Hesketh (ed.) Plant growth modeling for resource management, Vol. II: Quantifying plant processes, CRC Press, Boca Raton, FL.

Gutschick, V.P., and G.L. Cunningham. 1989. A physiological route to increased water-use efficiency in alfalfa. Tech. Completion Rep. 1423693. New Mexico Water Resour. Res. Inst., Las Cruces.

Gutschick, V.P., and F.W. Wiegel. 1988. Optimizing the canopy photosynthetic rate by patterns of investment in specific leaf mass. Am. Nat. 132:67–86.

Henson, I.E., C.R. Jensen, and N.C. Turner. 1989. Leaf gas exchange and water relations of lupins and wheat. I. Shoot responses to soil water deficits. Aust. J. Plant Physiol. 16:401–413.

Hiebsch, C.K., E.T. Kanemasu, and C.D. Nickell. 1976. Effects of soybean leaflet type on net carbon dioxide exchange, water use, and water-use efficiency. Can. J. Plant Sci. 56:455–458.

Hubick, K.T., G.D. Farquhar, and R. Shorter. 1986. Correlation between water-use efficiency and carbon isotope discrimination in diverse peanut (*Arachis*) germ-plasm. Aust. J. Plant Physiol. 13:803–816.

Jarvis, P.G., and K.G. McNaughton. 1986. Stomatal control of transpiration: Scaling up from leaf to region. Adv. Ecol. Res. 15:1–49.

Johnson, D.A., K.H. Assay, L.L. Tieszen, J.R. Ehleringer, and P.G. Jefferson. 1988. Carbon isotope discrimination as a possible selection criteria for improving range grass production in water-limited environments. p. 112. *In* Agronomy abstracts. ASA, Madison, WI.

Johnson, I.R., and J.H.M. Thornley. 1984. A model of instantaneous and daily canopy photosynthesis. J. Theor. Biol. 107:531–545.

Jones, H.G. 1981. The use of stochastic modelling to study the influence of stomatal behavior on yield–climate relationships. p. 231–244. *In* D.A. Rose and D.A. Charles-Edwards (ed.) Mathematics and plant physiology. Academic Press, London.

Jones, H.G. 1983. Plants and microclimate: A quantitative approach. Cambridge Univ. Press, Cambridge, England.

Kallis, A., and H. Tooming. 1974. Estimation of the influences of leaf photosynthetic parameters, specific leaf weight and growth functions on yield. Photosynthetica 8:91–103.

Küppers, M., R. Matyssel, and E.-D. Schulze. 1986. Diurnal variations of light-saturated CO_2 assimilation and intercellular carbon dioxide concentration are not related to leaf water potential. Oecologia 69:477–480.

Loomis, R.S., and W.A. Williams. 1969. Productivity and the morphology of crop stands: Patterns with leaves. p. 27–51. *In* J.D. Eastin et al. (ed.) Physiological aspects of crop yield. ASA and CSSA, Madison, WI.

Lösch, R., and J.D. Tenhunen. 1981. Stomatal responses to humidity: phenomenon and mechanism. p. 137–162. *In* P.G. Jarvis and T.A. Mansfield (ed.) Stomatal physiology. Cambridge Univ., Cambridge, England.

Mansfield, T.A., and W.J. Davies. 1985. Mechanisms for leaf control of gas exchange. BioScience 35:158–164.

Martin, B., J. Nienhuis, G. King, and A. Schaefer. 1989. Restriction fragment length polymorphisms associated with water use efficiency in tomato. Science (Washington, DC) 243:1725–1728.

Monsi, M., and T. Saeki. 1953. Über den Lichtfaktor in den Pflanzengesellschaften und seine Bedeutung für die Stoffproduktion. Jpn. J. Bot. 24:22–52.

Nelson, C.J. 1988. Genetic associations between photosynthetic characteristics and yield. Plant Physiol. Biochem. 26:543–554.

Pearce, R.B., G.E. Carlson, D.K. Barnes, R.H. Hart, and C.H. Hanson. 1969. Specific leaf weight and photosynthesis in alfalfa. Crop Sci. 9:423–426.

Penning de Vries, F.W.T. 1975. The cost of maintenance processes in plant cells. Ann. Bot. (London) 39:77–92.

Penning de Vries, F.W.T., A.H.M. Brunsting, and H.H. van Laar. 1974. Products, requirements, and efficiency of biosynthesis: a quantitative approach. J. Theor. Biol. 45:339–377.

Pushnik, J.C., B.A. Swanton, and V.P. Gutschick. 1988. Plant growth chambers at high irradiance. BioScience 38:44–47.

Reed, R., and R.L. Travis. 1987. Paraheliotropic leaf movements in mature alfalfa canopies. Crop Sci. 27:301–304.

Ross, J. 1981. The radiation regime and architecture of plant stands. Junk, The Hague.

Scott, D., and J.S. Wells. 1969. Leaf orientation in barley, lupin, and lucerne stands. N.Z. J. Bot. 7:372–388.

Sinclair, T.R., C.B. Tanner, and J.M. Bennett. 1984. Water-use efficiency in crop production. BioScience 34:36–40.

Song, S.P., and P.D. Walton. 1975. Inheritance of leaflet size and specific leaf weight in alfalfa. Crop Sci. 15:649–652.

Tanksley, S.D., and T.J. Orton. 1983. Isozymes in plant genetics and breeding. Elsevier, New York.

Thomas, H. 1986. Water use characteristics of *Dactylis glomerata* L., *Lolium perenne* L., and *L. multiflorum* Lam. plants. Ann. Bot. (London) 57:21–223.

Travis, R.L., and R. Reed. 1983. The solar tracking pattern in a closed alfalfa canopy. Crop Sci. 23:664–668.

Trenbath, B.R., and J.F. Angus. 1975. Leaf inclination and crop production. Field Crop Abstr. 28:231–244.

Versteeg, M.N., I. Zipori, J. Medina, and H. Valdiva. 1982. Potential growth of alfalfa (*Medicago sativa* L.) in the desert of southern Peru and its response to high NPK fertilization. Plant Soil 67:157–165.

Vos, J., and P.J. Oyarzún. 1987. Photosynthesis and stomatal conductance of potato leaves—Effects of leaf age, irradiance, and leaf water potential. Photosynth. Res. 11:253–264.

Wallace, D.H., J.L. Ozbun, and H.M. Munger. 1972. Physiological genetics of crop yield. Adv. Agron. 24:97–146.

Wilson, M., B. Melton, J. Arledge, D. Baltensperger, R.M. Salter, and C. Edminster. 1983. Performance of alfalfa cultivars under less than optimum moisture conditions. New Mexico State Univ. Agric. Exp. Stn. Bull. 702.

Wisiol, K., and J.D. Hesketh (ed.). 1987. Plant growth modeling for resource management. CRC Press, Boca Raton, FL.

Wong, S.-C., I.R. Cowan, and G.D. Farquhar. 1985a. Leaf conductance in relation to rate of CO_2 assimilation. 1. Influence of nitrogen nutrition, phosophorus nutrition, photon flux density, and ambient partial pressure of CO_2 during ontogeny. Plant Physiol. 78:821–825.

Wong, S.-C., I.R. Cowan, and G.D. Farquhar. 1985b. Leaf conductance in relation to rate of assimilation. 1. Effects of short-term exposure to different photon flux densities. Plant Physiol. 78:826–829.

Wong, S.-C., I.R. Cowan, and G.D. Farquhar. 1985c. Leaf conductance in relation to rate of CO_2 assimilation. 1. Influences of water stress and photoinhibition. Plant Physiol. 78:830–834.

5 Predicting Canopy Photosynthesis and Light-Use Efficiency from Leaf Characteristics

J. M. Norman

University of Wisconsin
Madison, WI

T. J. Arkebauer

University of Nebraska
Lincoln, NE

Vegetation depends on light energy from the sun for the conversion of CO_2 from the air to essential life-sustaining C compounds. Light has long been recognized as an essential factor in photosynthesis, which provides biochemical energy and C necessary for growth. The light-gathering system consists mainly of an aerial array of leaves, with a light-trapping effectiveness that depends on many plant and environmental factors. The operation of this light-gathering system is a key determinant of plant productivity, and a clearer understanding of it will help us to predict the growth of crops with more reliability. This chapter is taken largely from Norman and Arkebauer (1991) with some additional discussion of photosynthesis.

The accumulated growth of any plant depends on the total C fixed by photosynthesis and the fraction of that C that can be converted to dry matter. Although other nutrients than C also are essential to tissue growth, this chapter will consider only C. Only a portion of the C fixed by photosynthesis eventually appears as standing dry matter, because the plant respires away some of the C in two ways: synthesis of compounds that form the final dry matter, and maintenance of the "living complex" in a functioning condition.

A simplistic view of the disposition of C in a growing plant can be considered as a combination of three processes: photosynthetic C fixation, maintenance respiration, and growth respiration. Although considerable uncertainty exists in our knowledge of all three processes, the work of Penning de Vries et al. (1974) has increased our confidence in growth-respiration estimates so that C fixation and maintenance respiration represent greater concentrations of ignorance.

 Modeling Crop Photosynthesis—from Biochemistry to Canopy. CSSA Special Publication no. 19.

In this chapter, we will consider the photosynthetic C fixation and its efficiency in terms of the mass of CO_2 fixed per unit of absorbed photosynthetically active radiation; something called the photosynthetic light-use efficiency (LUE_p). The LUE_p of a canopy is analogous to the photochemical efficiency of a leaf. Alternatively, the mass of dry matter produced per unit of absorbed photosynthetically active radiation could be termed the dry-matter light-use efficiency (LUE_{dm}). Clearly LUE_{dm} involves maintenance and growth respiration, which may not depend on light directly, and photosynthesis, which directly depends on light interception.

The concept of a relatively constant LUE_{dm} has great potential for simplifying the prediction of plant productivity. Since incoming solar radiation (SR) or photosynthetically active radiation (PAR) are relatively easy to measure, the simple product of intercepted or absorbed PAR and the appropriate canopy LUE_{dm} could provide an estimate of dry-matter increment (Hesketh & Baker, 1967). This concept was used by Monteith (1977) to study the effect of climate on crop production in Britain. Monteith's approach was expanded by Charles-Edwards (1981), who carried out a very similar analysis. Numerous investigators have used this approach and measured the canopy LUE_{dm} for different crops. Monteith (1977) suggested a seasonal canopy LUE_{dm} value of 1.4 g dry matter (DM) MJ^{-1} intercepted solar radiation (ISR) based on aboveground data from apple (*Malus domestica* Borkh.), barley (*Hordeum vulgare* L.), sugar beet [*Beta vulgaris* ssp. *cicla* (L.) Koch], and potato (*Solanum tuberosum* L.). Gallagher and Biscoe (1978) found a seasonal value of 3.0 g DM MJ^{-1} absorbed photosynthetically active radiation (APAR) for wheat (*Triticum aestivum* L.) and barley in Britain including roots and tops; this is approximately equivalent to the value used by Monteith (1981) of 4.3 g CO_2 MJ^{-1} ISR as well as the value of 1.4 g DM MJ^{-1} ISR reported by Monteith (1977). Unsworth et al. (1984) measured a value of about 1.2 g DM MJ^{-1} ISR on soybean [*Glycine max* (L.) Merr.] at various moderate levels of ozone treatment. Muchow and Coates (1987) determined seasonal sorghum [*Sorghum bicolor* (L.) Moench] light-use efficiencies between 2.1 and 2.4 g DM MJ^{-1} IPAR for intercepted PAR on a seasonally integrated basis. Charles-Edwards (1981) has summarized some values of canopy light-use efficiency from the literature and simply reported them as g DM MJ^{-1}: rice (*Oryza sativa* L.), 4.2; corn (*Zea mays* L.), 3.4; sweet potato [*Ipomoea batatas* (L.) Lam.], 3.1; kale (*Brassica oleracea* var. *acephela* DC.), 2.7; sunflower (*Helianthus annuus* L.), 2.6; cotton (*Gossypium hirsutum* L.), 2.5; clover (*Trifolium* spp.), 1.6; and soybean, 1.3. Unfortunately, these values are difficult to interpret, since the form of the radiation is not indicated (solar or PAR).

The values of light-use efficiency reported in the literature vary by more than a factor of three. Two reasons for this are that (i) the values reported do not have a common basis, and (ii) differences in canopy photosynthesis and respiration occur among various crops and under different environmental conditions. The main objective of this chapter is to investigate a basis for a canopy light-use efficiency by using physiological characteristics of leaves and integrating to the canopy level with considerations of canopy architecture and environmental gradients.

The common basis for light-use-efficiency comparison in this chapter is grams of CO_2 megajoule^{-1} IPAR. Although the conversion of daily integrated values to other units (absorbed solar radiation [ASR], APAR, and ISR) depends on leaf area index (LAI), 8.0 g CO_2 MJ^{-1} IPAR is equivalent to 8.3 ± 0.1 g CO_2 MJ^{-1} APAR, 4.5 ± 0.4 g CO_2 MJ^{-1} ISR, and 5.7 ± 0.4 g CO_2 MJ^{-1} ASR, where the uncertainty factor represents the effect of LAI (Norman & Arkebauer, 1991). This value of 8.0 g CO_2 MJ^{-1} IPAR is approximately equivalent to 3.6 g DM MJ^{-1} IPAR for a crop like corn.

STAND LIGHT-USE EFFICIENCY DEFINITIONS

Throughout the literature, many definitions are used for canopy light-use efficiency; no standard form is apparent. A definition of canopy light-use efficiency involves three aspects: (i) the time interval—instantaneous, hourly, daily, weekly, or seasonal; (ii) the form of the C—dry matter above ground (DMAG), total plant dry matter including roots (DMT), or net CO_2 uptake by the plant top (CO_2); (iii) characterization of the radiation—ISR, ASR, IPAR, or APAR. Intercepted radiation is the incident radiation (I in MJ m^{-2}) minus the radiation transmitted (T in MJ m^{-2}) to the bottom of the canopy:

$$\text{IPAR} = I - T \qquad [1]$$

Absorbed radiation is the net radiation (PAR or SR) above the canopy (downward [I] minus upward [RC] minus the net radiation below the canopy (radiation transmitted [T] minus radiation reflected from the soil [RS]).

$$\text{APAR} = (I - \text{RC}) - (T - \text{RS}) \qquad [2]$$

The units in Eq. [1] and [2] are MJ m^{-2} in the appropriate wave band (PAR or SR). In the PAR waveband, 1 MJ m^{-2} is 4.6 mol quanta m^{-2} if the sun is the source of radiation.

The form of the C fixed affects the numerical values of canopy light-use efficiency. The conversion between grams of CO_2 and grams of DM depends on the composition of the plant, because the amount of C required to build carbohydrates, proteins, and lipids is different; furthermore, the amount of these constituents in various plants is different. Appropriate conversion factors can be obtained from Penning de Vries et al. (1974) or McDermitt and Loomis (1981). When dry matter is used to express the C fixed, sometimes the roots are included and sometimes they are not. Since the fraction of the total plant dry matter that is in roots may vary from 10% for some agronomic crops to 80% for some native prairie grasses, this is not a minor inconsistency.

In addition, roots may lose significant amounts of C by sloughing, exudation, and respiration. Furthermore, with annuals, the amount of dry matter standing is known to be derived from C fixed in a given growing season. With perennials, the standing biomass may have been accumulated over several years and the contribution of a given season difficult to quantify, especially if roots are a major component of carryover between growing seasons. These uncertainties in respiration and root losses make it essential to obtain good estimates of the C fixed by photosynthesis if we are to advance our knowledge of the C budget.

Muchow and Coates (1987) discussed the likely effect of including root dry matter and intercepted vs. absorbed SR and PAR in cereals. The various definitions of light-use efficiency can result in numerical values that differ by more than a factor of two for a given canopy and condition. This chapter contains a discussion of canopy light-use efficiency from the perspective of an integrative model that uses leaf characteristics and predicts canopy characteristics.

MODEL DESCRIPTIONS

The approach used in this chapter is to describe the photosynthetic and respiration response functions of individual leaves in various parts of a canopy and then to integrate their various contributions to obtain the canopy net photosynthetic rate over some appropriate time interval. This approach requires that we first describe the dependence of leaf photosynthesis and respiration on various factors; for example light, temperature, water status, leaf spectral properties, and position in the canopy. Canopy C exchange rates are estimated from a model entitled Cupid by combining the equations that describe leaf C-exchange rates with a characterization of canopy architecture, boundary measurements of ambient environment above the canopy and below the root zone, and equations that describe the convective, conductive, and radiative exchange processes throughout the soil–plant–atmosphere system. A description of canopy architecture includes the vertical distribution of stem and leaf area, leaf angle distributions, canopy height, and some information about the horizontal distribution of foliage such as random or clumped. Ambient environmental boundary conditions are obtained from measurements of air temperature, humidity, wind speed, solar radiation, and precipitation at a few meters above the canopy, and soil temperature and water content near the bottom of the root zone (0.5–2-m depth). The soil measurements can be made once per day or even less frequently for deeper depths because these conditions change slowly. However, boundary conditions above the canopy should be measured for every time step desired for the model (10 min–3 hr). Gradients throughout the soil–plant–atmosphere system are accommodated by simultaneously solving the convective, conductive, and radiative exchange equations with the equations that describe canopy architecture, physiological characteristics of the vegetation and soil properties. Because of the detailed treatment of radiative exchange, estimates of canopy LUE_p can be obtained from Cupid using intercepted or absorbed SR or PAR.

The stand LUE_{dm} can be obtained by combining the C fixed in photosynthesis with estimates of growth and maintenance respiration.

Leaf Model

The model that we use to predict leaf net photosynthetic rate is based on the work of von Caemmerer and Farquhar (1981). The C_3 leaf model equations used here are very similar to CULEAF, which is described with program listing in Norman (1986), except that the equation for describing the dependence of assimilation rate on temperature was obtained from Schoolfield et al. (1981), and the CO_2 compensation point in the absence of mitochondrial respiration (Γ^*) is calculated with the equation given in Brooks and Farquhar (1985). The basis of this model is the relation between net assimilation rate (A, μmol m^{-2} s^{-1}) and internal CO_2 concentration (C_i, μmol mol^{-1}) (von Caemmerer & Farquhar, 1981). The strength of this approach is that limitations to CO_2 uptake that occur in the diffusion path (primarily stomata) can be distinguished from limitations caused by the biochemistry of photosynthesis. Sample A vs. C_i data for soybean (C_3) and sorghum (C_4) leaves, which were obtained in the field under a range of incident light conditions, are given in Fig. 5-1. The leaf-photosynthesis submodel of the Cupid model requires mathematical equations that capture the essential characteristics of the dependence of leaf assimilation rate on all the major environmental factors (Norman, 1986). Not only is absorbed light important as shown in Fig. 5-1, but also the effects of leaf temperature, air vapor pressure, and plant water status. Measurements of these leaf responses on plants grown in controlled environments usually are not appropriate for estimating canopy photosynthesis of field-grown crops. Therefore, leaf gas-exchange measurements are made in the field with portable gas-exchange instruments.

The leaf response functions for photosynthesis that are used in this chapter are typical of C_4 (corn or sorghum) and C_3 (soybean or wheat) crops and are shown in Fig. 5-2. The dependence of assimilation rates on C_i and light from this model is shown in Fig. 5-2a for corn. Since the von Caemmerer and Farquhar (1981) leaf-photosynthesis model is appropriate for C_3 plants, the parameter values cannot be interpreted in the same way for a C_4 plant. However, the parameters of the von Caemmerer and Farquhar (1981) model were adjusted to fit corn photosynthetic and respiratory responses based on our own field measurements and those of Chmora and Oya (1967), Edmeades and Daynard (1979), and Vietor et al. (1977). Figure 5-2b contains a light-photosynthesis response for upper sunlit leaves of C_3 and C_4 crop types. The C_3 parameters are essentially those of von Caemmerer and Farquhar (1981) except for the temperature response mentioned above, and the maximum rate of electron transfer, maximum rate of carboxylation, and dark respiration were increased over the value they used so that maximum rates of photosynthesis in full sunlight are about

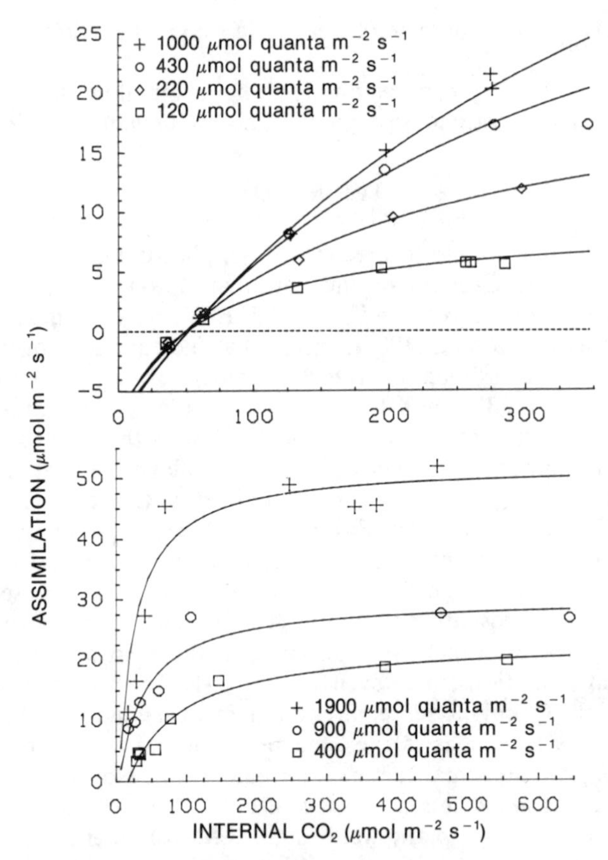

Fig. 5-1. The measured dependence of photosynthetic rate on internal CO_2 concentration for soybean (upper) and sorghum (lower) for a range of PAR flux densities. The sorghum measurements were made with an LI-6000 and the soybean with an LI-6200 (both from LI-COR, Lincoln, NE).

50% greater (see Fig. 5-2b). The main changes in the parameters of the C_3 model to simulate responses of C_4 photosynthesis are in the maximum rate and light dependence of electron transfer, the maximum rate of carboxylation, dark respiration, photochemical efficiency in low light (0.062), and the CO_2 compensation point (fixed at 2 μmol mol^{-1}).

The variation of leaf photosynthetic characteristics with depth in the canopy is an important factor to consider in the prediction of canopy photosynthesis from leaf photosynthesis, and enters into the Cupid model through the effects of a leaf's previous light history. The effect of previous light history of a leaf, which affects the maximum electron-transfer rate in full sun, the maximum rate of carboxylation, and dark respiration, is related to the LAI above the particular leaf. Leaves low in the canopy have lower light-saturated photosynthetic rates and higher photosynthetic rates in the shade than the uppermost leaves (Fig. 5-2b).

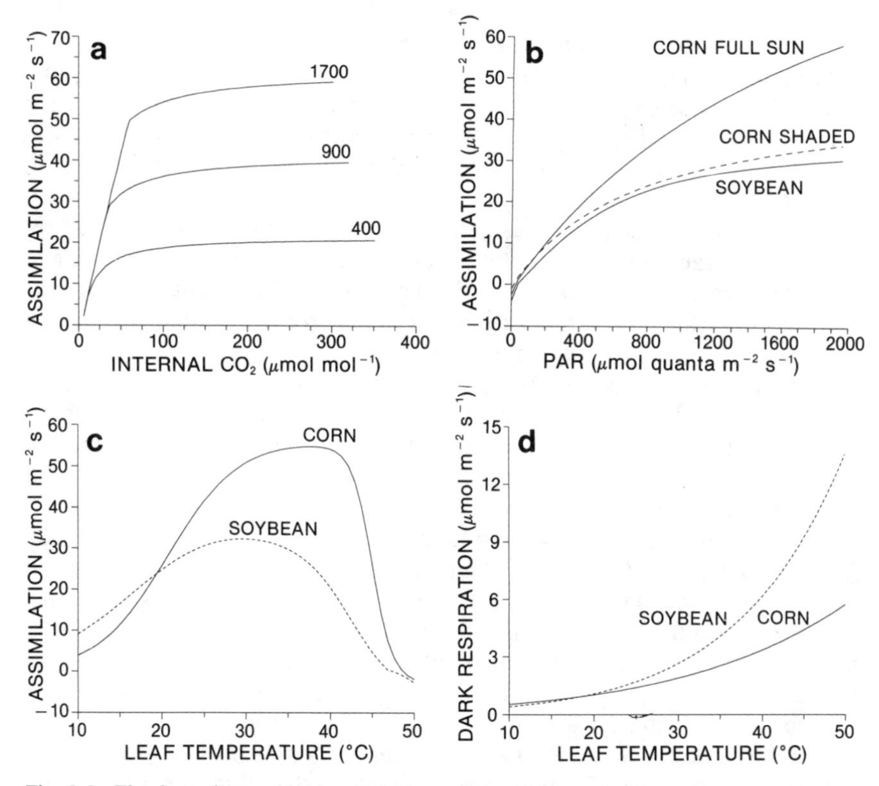

Fig. 5-2. The dependence of leaf assimilation rate on (a) internal CO_2 concentration for corn at three light levels (μmol quanta m^{-2} s^{-1}), (b) light for upper and lower corn leaves and an upper soybean leaf, (c) temperature for corn and soybean, and (d) the dependence of dark respiration on temperature for corn and soybean.

A typical dependence of leaf photosynthetic rate on temperature is shown in Fig. 5-2c for C_4 and C_3 plants. This type of response curve is difficult to obtain and probably changes as a plant adapts to varying temperature environments (de Wit, 1978). The measurements on corn of Chmora and Oya (1967) indicate moderate leaf photosynthetic rates at leaf temperatures above 50 °C. This is higher than upper limits measured in the field in Nebraska, but high temperature and leaf water stress are difficult to separate under field conditions. The simulations done for this chapter did not result in leaf temperatures above 35 °C, so uncertainties in the temperature curve above 40 °C do not affect these results. Curves such as that shown in Fig. 5-2c should not be taken too seriously for the purpose of predicting canopy photosynthesis from leaf photosynthesis. The curves may even vary with time of day (Chmora & Oya, 1967). They typically are obtained for a single leaf that will be experiencing a different temperature than the other nonenclosed leaves on the plant. Such leaves tend to show a more marked temperature dependence because of exposure to conditions to which they have not had time to acclimate. De Wit (1978) recognized this and, as part of his corn model, assumed that corn photosynthetic rate was independent of leaf temperature above 13 °C and linearly decreased to zero between 13 and 8 °C. This response is quite different from those of Chmora and Oya (1967) and Fig. 5–2c. In this chapter, we predict canopy light-use efficiency using the temperature dependence in Fig. 5–2c and compare the results with those obtained by assuming the photosynthetic response to temperature similar to de Wit (1978).

Stomatal conductance is calculated from leaf photosynthesis, leaf-to-air vapor-pressure deficit, and leaf water potential by a method similar to Norman (1986). An unstressed stomatal conductance is calculated from the leaf photosynthetic rate (from the von Caemmerer & Farquhar model) by assuming C_i/ambient CO_2 concentration (C_a) is constant. This ratio is known to vary by species and genotype, and we use 0.4 for C_4 crops and 0.72 for C_3 crops in this chapter. High leaf-to-air vapor-pressure deficits and low leaf water potentials can result in stomatal closure. For this chapter all simulations were done with adequate soil moisture. The dependence of stomatal conductance on vapor-pressure deficit is controversial for agricultural crops because its magnitude may depend on growing conditions. Controlled-environment-grown plants tend to have a greater dependence on vapor-pressure deficit than field-grown plants. The complications associated with predicting the effect of vapor-pressure deficit on stomatal conductance are formidable (Losch & Tenhunen, 1981; Schulze et al., 1987). Because of this uncertainty, C_i/C_a values listed above were chosen based on midday measurements from corn and soybean leaves when the vapor-pressure deficit was approximately 2 to 3 kPa. Choosing C_i/C_a in this way implicitly incorporates an effect of vapor-pressure deficit in the estimation of stomatal conductance from photosynthesis, and accomplishes reasonable agreement between predicted and measured conductances. A clever approach has been suggested by Ball et al. (1986), but it remains to be determined whether it reduces uncertainty in prediction of stomatal conductance under field conditions.

The leaf dark-respiration rate also depends on temperature and, in this chapter, we used a rate of 1.5 μmol m^{-2} s^{-1} at 25 °C for corn or sorghum with a Q_{10} of about 1.8 (i.e., for each 10 °C increase in temperature, respiration rate increases 1.8-fold), and 1.7 μmol m^{-2} s^{-1} at 25 °C for soybean or wheat with a Q_{10} of about 2.3 (Fig. 5–2d; 1987, unpublished data).

Leaf characteristics that are essential for the prediction of radiation interception or absorption are leaf reflectance and transmittance. Since light-use efficiency has been expressed in terms of both solar and photosynthetically active radiation, we will consider radiation extinction for both these wavelength bands. This is best accomplished by obtaining leaf spectral properties in visible and near-infrared (NIR) wavelength bands and solving the extinction and scattering equations separately for each of these bands. The light-extinction distributions in the PAR are used directly for predicting photosynthesis and stomatal conductance of individual leaves in various layers and leaf-angle classes. The sum of the photosynthetic rates of all the leaves then provides an estimate of the photosynthetic rate for the canopy. The

Table 5-1. Leaf reflectance and transmittance for corn and soybean in visible (PAR) and near-infrared (NIR) wavelength bands from field measurements on intact plants using an integrating sphere.

	Leaf reflectance		Leaf transmittance	
	PAR	NIR	PAR	NIR
	%			
Corn	9	38	4	45
Soybean	9	42	4	42

radiation-penetration results from the PAR and NIR wavelength bands are added together to obtain the ASR or ISR for energy-balance considerations. Table 5-1 contains some leaf spectral properties measured in the field on intact plants (Walter-Shea, 1987). These leaf spectral characteristics also are important for determining the fraction of incident quanta that are absorbed, permitting the photochemical efficiency to be used in the conversion of light to CO_2 fixed.

Canopy Model

The collective effect of all the leaves must be obtained to predict the canopy light use. The light, temperature, humidity, and wind environments appropriate for each leaf are obtained from a combined solution of the leaf energy budget (for leaves in various angle classes and layers) and vertical-profile equations for radiation and turbulent transfer (Norman, 1979; Norman & Campbell, 1983). An iterative-solution technique is used to solve the combination of vertical-profile equations for convective exchange (Norman & Campbell, 1983) and the leaf energy-budget equations (Norman, 1979). The fluxes of CO_2, water vapor, energy, and radiation from leaves in various leaf-angle classes are summed to obtain the layer source or sink necessary for the solution of the vertical profile equations. The temperature, humidity, etc. from the solution to the vertical-profile equations are necessary to the solution of the leaf energy-budget equations. Of course, solution of a leaf energy-budget equation for leaves in each angle class required a stomatal conductance for each angle class. Because photosynthetic rate is required to calculate stomatal conductance, the Farquhar and von Caemmerer (1982) model becomes a part of this iteration loop.

This methodology incorporates aerodynamic resistances above and within the canopy, and boundary-layer resistances of individual leaves. Because of these additional resistances in the leaf-to-atmosphere path for water vapor, stomatal conductance may exhibit a lesser control over canopy transpiration than over leaf transpiration until severe closure occurs, especially in canopies shorter than 1 or 2 m.

The Cupid model, which is used for the simulations reported in this chapter, incorporates many plant, atmospheric, and soil processes with considerable detail in an attempt to provide a defendable integration from the leaf level to the canopy level (Norman, 1979, 1986; Norman & Campbell, 1983). However, some processes that may affect canopy light-use efficiency have not been considered. For example, the fraction of light absorbed by leaves may depend on the incidence angle of the direct beam on the leaf (Walter-Shea, 1987). We have assumed leaves to be Lambertian scatterers and thus may slightly overestimate photosynthesis of some leaves. We also have ignored the fluctuating light conditions within the canopy. High-frequency light fluctuations from wind effects may increase photosynthetic rates, and low-frequency fluctuations from sun angle shifts may decrease photosynthetic rates because of stomatal and photosynthetic-induction delays (Pearcy, 1988). Further, direct effects of wind on stomata and the leaf-angle distribution have not been considered, nor have effects of leaf movement. In fact, the only leaf-angle distribution used for these simulations is the spherical distribution, which is reasonable for corn and vegetative stages of soybean growth.

The radiation-extinction equations of Norman (1979), which are for random leaf positioning, can be modified for the clumping effect of rows by using the clumping factor defined by Nilson (1971). This effect improves light-interception estimates for canopies of partial cover, but appears to be minor in crops such as corn or soybeans as full cover is approached.

Respiration Considerations

The instantaneous net CO_2 taken up by a plant is the difference between CO_2 fixed by photosynthesis and that respired by growth and maintenance respiration processes. The net photosynthetic rate measured with leaf chambers in the field provides us with an estimate of the net CO_2 exchange for leaves. However, other organs, such as leaf sheaths, roots, ears, and tassels, enter into considerations of C exchange. The distinction between photosynthesis and respiration processes is clear in terms of substrates but becomes blurred in terms of biochemical energy supply, since both processes may be occurring in the same leaf cells. Since respiration in the light may be different from respiration in the dark (Sharp et al., 1984; Brooks & Farquhar, 1985), the approach of de Wit (1978) seems desirable—that is to use CO_2-exchange measurements during day and night hours to estimate net C exchange of leaves and consider the growth and maintenance requirements of nonphotosynthesizing organs separately. Another approach may be to estimate gross photosynthesis of leaves and then consider leaf growth and maintenance requirements by similar methods as used for nonphotosynthesizing organs but perhaps with different coefficients (Baker et al., 1972). Measurements of stand photosynthesis with canopy enclosures often are confounded by soil CO_2 evolution, so measurements in the light are followed immediately with measurements in a darkened chamber to obtain an estimate of gross canopy photosynthetic rate. This is valid to the extent that respiration in the dark is equal to respiration in the light. Obviously, a perfect separation of photosynthesis and respiration gas exchanges is very

Table 5-2. Typical characteristics of a corn stand just before tasseling in the Corn Belt, USA.

Characteristic	Value
Planting density	6 plants m^{-2}
Height	2.0 m
Row spacing	0.76 m
Leaf area index	3.5
Leaf dry wt./leaf area	77 g m^{-2}
Dry wt./ground area:	
Total vegetative	900 g m^{-2}
Leaves	270 g m^{-2}
Sheaths	135 g m^{-2}
Roots	90 g m^{-2}
Stems	405 g m^{-2}

difficult and the best methods may depend on how one plans to verify any particular model with measurments.

The consideration of the detailed C budget for an entire season, such as McCree (1988) did for sorghum, is beyond the scope of this chapter; therefore, let us compute canopy assimilation for a corn plant just prior to tasseling so that vegetative components dominate. Based on data for corn from Foth (1962), de Wit (1978), Yao (1980), and Righes (1980), a typical stand just prior to tasseling in the U.S. Corn Belt might have the characteristics shown in Table 5-2. The final seasonal grain yield for this typical corn stand might be 1 kg m^{-2} (175 bu/acre) and, at the end of the season, this yield would be about one-half of the total accumulated dry matter.

For the purpose of this chapter, we will assume that corn leaves are not growing, which is valid just prior to tasseling, and we will further assume that the net photosynthesis and nighttime respiration relations given in Fig. 5-2 will represent the net C source provided by a leaf. Thus the C cost of phloem loading, various C transformations, and maintenance respiration of leaves during the day are accommodated in the net photosynthesis relations, and maintenance respiration at night is accounted for by the dark-respiration vs. temperature relation. At 25 °C, the daily dark respiration for the leaf respiration rate of 1.4 μmol m^{-2} s^{-1} assumed for corn compares with the daily maintenance coefficient of 0.02 g glucose g^{-1} leaf dry wt. d^{-1} for the stand characteriztics summarized in Table 5-2. This is slightly less than the value of 0.03 g glucose g^{-1} dry wt d^{-1} given by McCree (1988) for sorghum leaves.

The maintenance requirements of leaf sheaths are assumed to be offset by their photosynthesis, the maintenance respiration of the stem is assumed small compared with that of leaves and taken as one-tenth the leaf value, and the maintenance respiration of roots is assumed to be one-third that of leaves (McCree, 1988) because of sloughing, exudation, and ion exchange (Amthor, 1984; Lambers, 1985, 1987). Much uncertainty exists about C lost in the root system. Measurements by Andre et al. (1978) of root respiration plus exudation in a controlled-environment experiment are consistent with a maintenance coefficient for roots about equal to that for leaves. Furthermore, these root-maintenance costs decrease during grain fill. McCree (1988) used a stem-maintenance coefficient in sorghum of one-half that of the leaf, but he suggested that stem-maintenance coefficients in sorghum were much lower (McCree, 1988, personal communication). Thus, at 25 °C, the maintenance respiration of roots and stems just before tasseling would be 0.0067 and 0.002 g glucose g^{-1} dry wt. d^{-1}, respectively, with a Q_{10} of 2.0.

The growth respiration can be computed from the work of Penning de Vries based on biochemical pathways using a constituent composition typical of corn (Vertregt & Penning de Vries, 1987). The cost of converting glucose to dry matter, termed the production value inverse (PVI) by Vertregt and Penning de Vries (1987), permits an estimate of the dry-matter increment that could be realized from a given amount of net photosynthesis that has had maintenance costs subtracted. For corn, the PVI is 1.42 g glucose g^{-1} dry wt. (Vertregt & Penning de Vries, 1987).

MODEL COMPARISONS WITH MEASUREMENTS

The radiation model that is used in this chapter is based on several assumptions: random leaf positioning in the horizontal plane, azimuthal symmetry, and a well-defined leaf-inclination distribution such as spherical, conical, or vertical. This kind of model has been compared with measurements in several crops of full cover and shown to perform well in both PAR (Norman et al., 1971; Norman, 1988) and NIR (Norman et al., 1971) wavelength bands.

During the summer of 1981, measurements of IPAR and canopy photosynthesis were made in corn stands of various planting densities near Oneil, NE, through a combined effort of Joe Ritchie (Michigan State Univ., East Lansing), Dan Knievel (Pennsylvania State Univ., University Park), Don Reicosky (Univ. of Minnesota, Morris), and John Norman (Univ. of Nebraska, Lincoln). The planting densities varied from 2.3 to 7.4 plants m^{-2}, the LAI varied from about 1.3 to 3.7, and the date of measurement was 4 August. The measured light-use efficiency was 6.9 g CO_2 MJ^{-1} IPAR. This light-use efficiency is in good agreement with the results of the Cupid model, which averaged 6.8 g CO_2 MJ^{-1} IPAR over 17 instantaneous periods that corresponded to chamber measurements at the three LAIs of 1.3, 2.2, and 3.7. The comparison of canopy-photosynthesis predictions from Cupid with direct enclosure measurements (Reicosky & Peters, 1977; Reicosky, 1985) are shown in Fig. 5–3. While the agreement is reasonable, considerable uncertainty could exist in the soil-respiration corrections. Soil respiration was estimated from bare soil measurements to be 10 μmol CO_2 m^{-2} s^{-1} and this value was added to all chamber measurements to estimate canopy net photosynthetic rate. Some chamber measurements were done with a loose-fitting plastic sheet over the soil surface. Although this may have

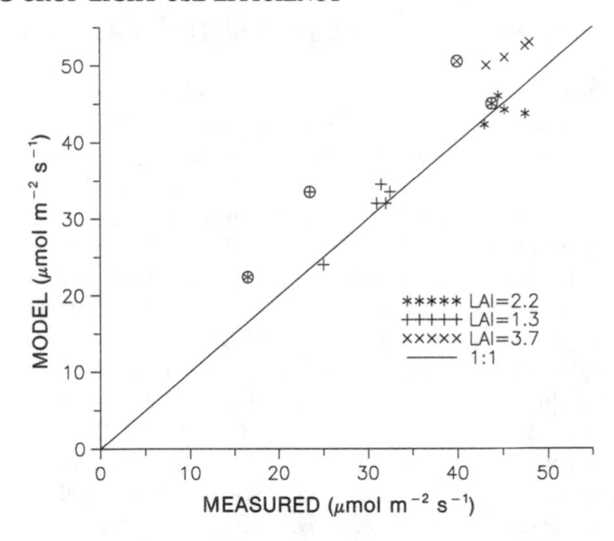

Fig. 5-3. Comparison of predicted canopy photosynthetic rates from O'Neill, NE, on 4 Aug. 1981 with chamber measurements. All chamber results have a soil respiration of 10 μmol m^{-2} s^{-1} added to provide an estimate of canopy net photosynthesis except circled values, which were measurements made with a loose-fitting plastic sheet covering the soil.

reduced soil respiration, the results in Fig. 5-3 suggest that it did not eliminate the soil CO_2 flux.

REVIEW OF SOME LIGHT-USE EFFICIENCY MEASUREMENTS

Measurements of PAR intercepted by a canopy and canopy photosynthesis can be used to estimate short-term (minutes to hours) canopy light-use efficiency. Care must be taken to eliminate soil respiration. Chambers can also be used to estimate light-use efficiency from CO_2 flux measurements over longer periods of days, weeks, or even months. Measurements of IPAR and dry matter over longer periods can be used to estimate LUE_{dm}. We have limited this discussion to canopies of full cover (LAI >3).

Jones et al. (1986) measured IPAR and canopy (including soil) C-exchange rates on corn at 50% silking and 6 d later. Assuming a soil respiration rate of 4 μmol CO_2 m^{-2} s^{-1}, at solar noon the light-use efficiency was 8.1 and 5.7 g CO_2 MJ^{-1} IPAR for the two respective days. These values are near predictions from Cupid given below.

Jones et al. (1985a) measured soybean canopy light-use efficiencies in outdoor controlled-environment chambers at growth stages R4 and R5 about 60 to 80 d after planting. The average daytime light-use efficiencies were about 6.0 g CO_2 MJ^{-1} PAR for an LAI of 3.3. Jones et al. (1985b) determined a seasonal light-use efficiency of about 5.0 g CO_2 MJ^{-1} IPAR on soybeans. Jones et al. (1987, unpublished data) provided data on tomato [*Lycopersicon lycopersicum* (L.) Karsten] for plants grown in outdoor chambers at two nighttime temperatures. The light-use efficiency was 6.2 g CO_2 MJ^{-1} IPAR for a 20 °C night temperature and 5.6 g CO_2 MJ^{-1} IPAR for a 12 °C night temperature.

The light-use efficiency was measured for a wheat canopy near midday about 40 to 70 d after sowing when the LAI was between 3.5 and 5.0 (Puckridge & Ratkowsky, 1971). After correcting for soil respiration, the light-use efficiency was 4.2 g CO_2 MJ^{-1} IPAR for IPAR of 1600 μmol quanta m^{-1} s^{-1}. At IPAR of 640 and 320 μmol quanta m^{-2} s^{-1}, the light-use efficiency was 6.3 and 7.1 g CO_2 MJ^{-1} IPAR. These values are similar to those reported for soybean. Spiertz and van deHaar (1978) also reported midday light-use efficiency of 4.7 g CO_2 MJ^{-1} IPAR for wheat about 1 wk after anthesis (23 June) with LAI $\approx$ 4 and IPAR = 1800 μmol quanta m^{-2} s^{-1}. Their light-use efficiency was 6.9 and 7.1 g CO_2 MJ^{-1} IPAR at 700 and 350 μmol quanta m^{-2} s^{-1} IPAR. The daily light-use efficiency from Spiertz and van deHaar (1978) was 4.7 g CO_2 MJ^{-1} IPAR for 23 June, which had a total ISR of 22 MJ m^{-2}. Over the week of 23 June, the measurements of Spiertz and van deHaar (1978) can be used to estimate an average light-use efficiency of 2.2 g DMAG MJ^{-1} IPAR; this dry-matter value is 47% of the daily light-use efficiency from CO_2 flux measurements.

Estimates of LUE_{dm} vary from 2.9 (Williams et al., 1965) to 3.2 (Yao, 1979) to 3.8 (Sivakumar & Virmani, 1984) to 4.4 g DMAG MJ^{-1} IPAR (Griffin, 1980) for corn grown under field conditions. This is a wide range of values and a clearer understanding of the reason for this range would be most useful. Light-use-efficiency estimates for sorghum are similar: 2.4 (Muchow & Coates, 1986), 2.9 (Sivakumar & Virmani, 1984), and 3.0 g DMAG MJ^{-1} IPAR (Steiner, 1986).

INTERPRETATIONS FROM THE CUPID MODEL

The plant–environment model Cupid provides a means for studying the characteristics of crop light-use efficiency. Using hourly SR and weather data, along with various soil and plant characteristics (Norman & Campbell, 1983), hourly canopy photosynthesis, and radiation penetration are predicted with Cupid.

The simulations from the model Cupid are based on 8 d of hourly SR and weather data collected over corn in 1981 at Garden City, KS, by J. Steiner and E.T. Kanemasu of Kansas State University. They also measured the LAI (2.8) and crop height (2 m). Table 5–3 contains a summary of the SR and weather data for the 8 d. These 8 d represent a wide range of radiation and temperature conditions typical of the Corn Belt in the central USA.

The light-use efficiencies of a C_4 crop (typical of corn) and a C_3 crop (typical of soybean) were calculated hour by hour with Cupid for the 8 d. The corn canopy was assumed to have a spherical leaf-angle distribution. We will use the symbolism

$$Q(CO_2, \text{IPAR}) = \text{light-use efficiency}$$

Table 5-3. Daily summary of solar radiation, air temperature, air vapor pressure, and mean wind speed for the 8 d used in simulations.

Day no.	Solar rad.	Air temperature Mean	Max.	Min.	Vapor pressure	Wind speed
	MJ	°C			mbar	m s^{-1}
201	29.7	27.5	35.0	18.2	22.5	2.0
202	29.6	28.0	37.5	21.2	23.2	3.2
203	14.1	22.1	26.1	18.4	22.4	2.6
204	23.4	26.6	33.4	19.9	23.2	2.0
205	19.5	27.6	35.5	22.3	22.4	2.5
206	21.0	23.7	29.9	17.8	22.7	1.7
207	15.3	21.9	25.5	19.2	23.9	2.3
208	16.9	21.8	27.8	17.8	22.0	2.2

where the quantities in parentheses refer to the basis for the efficiency definition. The quantity $Q(CO_2$, IPAR) includes daytime photosynthesis plus nighttime dark respiration. This is the sum over the canopy of what is measured with a leaf gas-exchange system. The hourly values of $Q(CO_2$, IPAR) varied from about 6 to 10 g CO_2 MJ^{-1} IPAR for corn and 4 to 12 g CO_2 MJ^{-1} IPAR for soybean (Fig. 5-4). The upper values for soybean, which occur only in the early morning hours, are close to its photochemical efficiency of about 12 g MJ^{-1} APAR. Clearly, many factors can affect both light penetration and canopy photosynthesis on an hourly basis. One factor that appears to be causing variation in canopy light-use efficiency is the fraction of radiation above the canopy that is in the form of a direct beam (Fig. 5-4). When the radiation above the canopy is mainly diffuse, the canopy light-use efficiency is larger. This occurs because diffuse light is more

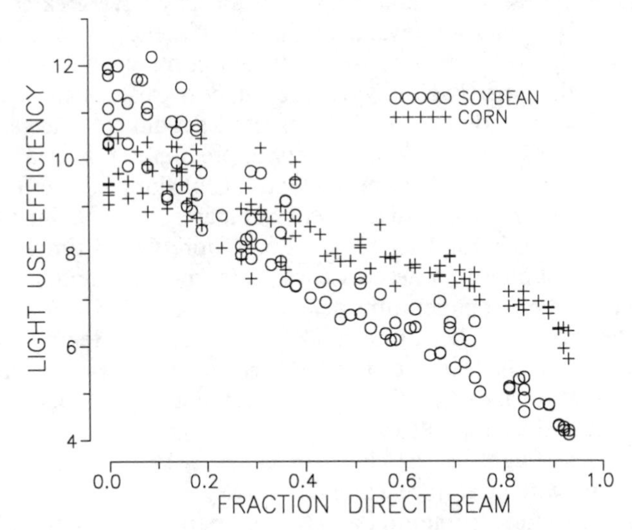

Fig. 5-4. The dependence of predicted light-use efficiency (in g CO_2 MJ^{-1} IPAR) on the fraction of PAR radiation above the canopy that is a direct beam for C_3 (soybean) and C_4 (corn) canopies.

Table 5-4. Predicted mean and standard deviation (in parentheses) of eight daily light-use efficiencies for typical C_3 and C_4 canopies at several leaf area indices (LAI) using the measured weather data summarized in Table 2.

Canopy	LAI	$Q(CO_2, IPAR)$	$Q(CO_2, APAR)$	$Q(CO_2, ISR)$	$Q(CO_2, ASR)$
		g CO_2 MJ^{-1}			
C_4	1	7.6 (0.4)	7.7 (0.4)	4.7 (0.2)	5.8 (0.3)
(corn)	3	7.5 (0.5)	7.8 (0.5)	4.2 (0.3)	5.3 (0.4)
	5	7.2 (0.5)	7.5 (0.5)	3.7 (0.2)	4.8 (0.3)
C_3	1	6.5 (0.8)	6.6 (0.9)	4.1 (0.5)	5.0 (0.7)
(soybean	3	6.5 (1.0)	6.7 (1.0)	3.6 (0.5)	4.6 (0.7)
or wheat)	5	6.2 (1.0)	6.5 (1.1)	3.2 (0.5)	4.2 (0.7)

uniformly and efficiently distributed over a canopy of leaves that may saturate at high light intensities. When the radiation incident on a canopy is diffuse, usually the illumination levels are low so that the quantity of intercepted radiation is low even though the light-use efficiency may be high. The daily light-use efficiency is calculated from the ratio of daily net photosynthesis and daily IPAR. Obviously, the daily light-use efficiency should not be calculated from an average of the hourly light-use efficiencies. The results in Fig. 5-4 incorporate many factors besides the fraction of beam radiation above the canopy (e.g., temperature, sun zenith angle, vapor-pressure deficit, and magnitude of the incident radiation). From various tests, however, the dominant factor influencing light-use efficiency is the fraction of diffuse radiation.

The daily light-use efficiencies do not vary as much as the hourly values do. The average of the eight daily light-use efficiencies for soybean are about 85% those for corn (Table 5-4). On overcast days, the light-use efficiencies of corn and soybean are similar and soybean may have greater efficiency on cool days. On clear, hot days, the light-use efficiency of soybean is about three-fourths that of corn. Apparently, the occurrence of two clear, three overcast, and three partly cloudy days resulted in greater light-use efficiencies for corn than for soybean. If leaf characteristics did not change, however, we would expect corn to be more efficient in predominantly clear, warm climates and soybean more efficient in cloudy, cool climates. Both C_3 and C_4 plant types adapt to their radiation environments, so such simple-minded statements may not hold. To simulate adaptation, Cupid simulations were run assuming that photosynthetic rate was independent of temperature (dark respiration remained temperature dependent) (Table 5-5). For C_3 simulations, the leaf photosynthetic rate at 30 °C was used at all temperatures, and for C_4 simulations the leaf photosynthetic rate at 35 °C was used regardless of leaf temperature above 15 °C. With this assumption, light-use efficiencies increased about 5 to 10% (Table 5-5).

The values of light-use efficiency in Tables 5-4 and 5-5 are in reasonable agreement with literature values discussed above based on CO_2 exchange. Establishing a common basis for comparison is not easy, however, because of some uncertainty in how maintenance and growth respiration should be accommodated. Because of these difficulties, we will only convert

Table 5–5. Mean and standard deviation (in parentheses) of eight daily light-use efficiencies predicted from the model Cupid for typical C_3 and C_4 canopies, assuming that leaf photosynthetic rate is independent of temperature. The same input data were used as in Table 5–4.

Canopy	LAI	$Q(CO_2$, IPAR)
		g CO_2 MJ^{-1}
C_4 (corn)	1	8.2 (0.5)
	3	8.0 (0.6)
	5	7.7 (0.6)
C_3 (soybean or wheat)	1	6.4 (0.7)
	3	6.3 (0.8)
	5	6.0 (0.8)

the CO_2 light-use efficiency to LUE_{dm} for the corn characteristics given in Table 5–2. The various components of the C balance of this corn canopy near tasseling are given in Table 5–6. The final LUE_{dm} of 3.2 g DM MJ^{-1} is near values measured for corn and summarized above; values between 3 and 4 g DM MJ^{-1} are typical.

Numerous factors are important in an exercise such as we have undertaken in this chapter (e.g., variations in leaf photosynthetic and respiratory responses with age and depth in the canopy, leaf area per unit leaf dry weight, partitioning in the plant, and maintenance-respiration coefficients).

SUMMARY

The plant–environment model Cupid is useful in studying canopy light-use efficiency by combining knowledge of leaf physiological and radiative

Table 5–6. Mean daily C exchanges for the canopy described in Table 5–2.

	Canopy C exchange
Photosynthesis	
Gross	73.7 (14.0) g CO_2 m^{-2} d^{-1}†
Gross minus leaf respiration	65.8 (12.7) g CO_2 m^{-2} d^{-1}
Net (gross minus total maintenance)	63.8 (12.5) g CO_2 m^{-2} d^{-1}
Maintenance respiration	
Stem	1.2 (0.2) g CO_2 m^{-2} d^{-1}
Leaf	7.9 (1.3) g CO_2 m^{-2} d^{-1}
Root	0.8 (0.1) g CO_2 m^{-2} d^{-1}
Grain	0
Total	9.9 (1.3) g CO_2 m^{-2} d^{-1}
Growth respiration	12.0 (2.3) g CO_2 m^{-2} d^{-1}
Dry-matter increment	29.2 (5.7) g DM m^{-2} d^{-1}‡
IPAR	9.0 (2.4) MJ m^{-2} d^{-1}
Light-use efficiency (CO_2, PAR)	7.1 g CO_2 MJ^{-1}
Light-use efficiency (DM, PAR)	3.2 g DM MJ^{-1}

† Values in parentheses are standard deviations for 8 d of simulation.
‡ 63.8 g CO_2 m^{-2} d^{-1} × 0.68 g glucose g^{-1} CO_2/1.49 g glucose g^{-1} DM = 29.2. The PVI of 1.49 g glucose g^{-1} DM includes 5% lost through processes other than biosynthesis (Vertregt & Penning de Vries, 1987, p. 118).

properties with canopy architecture. Hourly canopy light-use efficiencies vary more than daily values, especially for C_3 crops. Measured and modeled short-term (< 1 h) canopy light-use efficiencies agree well at values near 6.5 g CO_2 MJ^{-1} IPAR for corn. Predicted canopy photosynthetic rates agree within about 10% with chamber measurements in corn, but uncertainties in soil CO_2 efflux makes comparisons difficult. Even though the maximum, light-saturated, leaf photosynthetic rates used in the simulations reported in this chapter are high, the LUE_{dm} is not as large as some measured values in the literature.

The model Cupid has been used to predict conversion factors among intercepted or absorbed SR or PAR. These may be useful for resolving the many ways in which canopy light-use efficiencies are expressed in the literature.

The canopy light-use efficiency can vary with environmental conditions (radiation and temperature), LAI, and maintenance respiration. Maintenance-respiration coefficients are difficult to measure. Variations in maintenance respiration are not likely to be accommodated indirectly using only light-use efficiency to estimate dry-matter increment from intercepted light. The LUE_{dm} coefficients should be used with caution in simple models because of the potential for systematic errors. Furthermore, use of such coefficients may obscure possibilities for increased productivity by reductions in maintenance respiration.

REFERENCES

Amthor, J.S. 1984. The role of maintenance respiration in plant growth. Plant Cell Environ. 7:561–569.

Andre, M., D. Massimino, and A. Daguenet. 1978. Daily patterns under the life cycle of a maize crop. II. Mineral nutrition, root respiration and root excretion. Physiol. Plant. 44:197–204.

Baker, D.N., J.D. Hesketh, and W.G. Duncan. 1972. Simulation of growth and yield in cotton: I. Gross photosynthesis, respiration and growth. Crop Sci. 12:431–439.

Ball, J.T., I.E. Woodrow, and J.A. Berry. 1986. A model predicting stomatal conductance and its contribution to the control of photosynthesis under different environmental conditions. p. 221–224. *In* J. Biggins (ed.) Progress in photosynthesis research. Vol. 4. Martinus Nijhoff, Dordrecht, the Netherlands.

Brooks, A., and G.D. Farquhar. 1985. Effect of temperature on the CO_2/O_2 specificity of ribulose-1,5-bisphosphate carboxylase/oxygenase and the rate of respiration in the light. Planta 165:397–406.

Charles-Edwards, D.A. 1981. Physiological determinants of crop growth. Academic Press, New York.

Chmora, S.N., and V.M. Oya. 1967. Photosynthesis in leaves as a function of temperature. Sov. Plant Physiol. (Engl. Transl.) 14:513–519.

deWit, C.T. 1978. Simulation of assimilation, respiration and transpiration of crops. John Wiley & Sons, New York.

Edmeades, G.O., and T.B. Daynard. 1979. The relationship between yield and photosynthesis at flowering in individual maize plants. Can. J. Plant Sci. 59:585–601.

Farquhar, G.D., and S. von Caemmerer. 1982. Modeling of photosynthetic response to environmental conditions. p. 549–588. *In* O.L. Lange et al. (ed.) Physiological plant ecology II. New Ser. Vol. 12B. Encyclopedia of Plant Physiology. Springer-Verlag, Berlin.

Foth, H.D. 1962. Root and top growth of corn. Agron. J. 54:49–52.

Gallagher, J.N., and P.V. Biscoe. 1978. Radiation absorption, growth and yield of cereals. J. Agric. Sci. (Cambridge) 91:47–60.

Griffin, J.L. 1980. Quantification of the effects of water stress on corn growth and yield. M.S. thesis. Univ. of Missouri, Columbia.

Hesketh, J., and D. Baker. 1967. Light and carbon assimilation by plant communities. Crop Sci. 7:285–293.

Jones, J.W., B. Zur, and J.M. Bennett. 1986. Interactive effects of water and nitrogen stresses on carbon and water vapor exchange of corn canopies. Agric. For. Meteorol. 38:113–126.

Jones, P., L.H. Allen, Jr., and J.W. Jones. 1985a. Responses of soybean canopy photosynthesis and transpiration to whole-day temperature changes in different CO_2 environments. Agron. J. 77:242–249.

Jones, P., J.W. Jones, and L.H. Allen, Jr. 1985b. Seasonal carbon and water balances of soybeans grown under stress treatments in sunlit chambers. Trans. ASAE 28:2021–2028.

Lambers, H. 1985. Respiration in intact plants and tissues: Its regulation and dependence on environmental factors, metabolism and invaded organisms. p. 418–473. *In* R. Douce and D.A. Day (ed.) Encyclopedia of plant physiology. New Ser. Vol. 18. Springer-Verlag, Berlin.

Lambers, H. 1987. Growth, respiration, exudation and symbiotic associations: The fate of carbon translocated to the roots. p. 125–145. *In* P.J. Gregory et al. (ed.) Root development and function. Soc. Exp. Biol. Semin. Ser. 30. Cambridge Univ. Press, Cambridge, England.

Losch, R., and J.D. Tenhunen. 1981. Stomatal responses to humidity—phenomenon and mechanism. p. 137–161. *In* P.G. Jarvis and T.A. Mansfield (ed.) Stomatal physiology. Cambridge Univ. Press, Cambridge, England.

McCree, K.J. 1988. Sensitivity of sorghum grain yield to ontogenetic changes in respiration coefficients. Crop Sci. 28:114–120.

McDermitt, D.K., and R.S. Loomis. 1981. Elemental composition of biomass and its relation to energy content, growth efficiency and growth yield. Ann. Bot. (London) 48:275–290.

Monteith, J.L. 1977. Climate and the efficiency of crop production in Britain. Philos. Trans. R. Soc. London B 281:277–294.

Monteith, J.L. 1981. Climate variation and the growth of crops. Q.J.R. Meteorol. Soc. 107:749–774.

Muchow, R.C., and D.B. Coates. 1986. An analysis of the environmental limitations to yield of irrigated grain sorghum during the dry season in tropical Australia using a radiation interception model. Aust. J. Agric. Res. 37:135–148.

Nilson, T. 1971. A theoretical analysis of the frequency of gaps in plant stands. Agric. Meteorol. 8:25–38.

Norman, J.M. 1979. Modeling the complete crop canopy. p. 249–277. *In* B.J. Barfield and J. Gerber (ed.) Modification of the aerial environment of crops. ASAE, St. Joseph, MI.

Norman, J.M. 1986. Instrumentation use in a comprehensive description of plant–enviornment interactions. p. 149–307. *In* W. Gensler (ed.) Advanced agricultural instrumentation. Martinus Nijhoff Publ., The Hague.

Norman, J.M. 1988. Synthesis of canopy processes. p. 161–175. *In* G. Russell et al. (ed.) Plant canopies: Their growth, form and function. Cambridge Univ. Press, Cambridge, England.

Norman, J.M., and T.J. Arkebauer. 1991. Predicting canopy light-use efficiency from leaf characteristics. *In* J. Hanks and J.T. Ritchie (ed.) Modeling plant and soil systems. Agron. Monogr. 31. ASA, CSSA, and SSSA, Madison, WI.

Norman, J.M., and G.S. Campbell. 1983. Application of a plant-environment model to problems in irrigation. p. 155–188. *In* D.I. Hillel (ed.) Advances in irrigation. Academic Press, New York.

Norman, J.M., E.E. Miller, and C.B. Tanner. 1971. Light intensity and sunfleck size distributions in plant canopies. Agron. J. 63:743–748.

Pearcy, R.W. 1988. Photosynthetic utilization of light flecks by understory plants. Aust. J. Plant Physiol. 15:223–238.

Penning de Vries, F.W.T., A.H.M. Brunsting, and H.H. van Laar. 1974. Products, requirements and efficiency of biosynthesis: A quantitative approach. J. Theor. Biol. 45:339–377.

Puckridge, D.W., and D.A. Ratkowsky. 1971. Photosynthesis of wheat under field conditions: IV. The influence of density and leaf area index on the response to radiation. Aust. J. Agric. Res. 22:11–20.

Reicosky, D.C. 1985. Advances in evapotranspiration measured using portable field chambers. p. 79–86. *In* Advances in evapotranspiration. Proc. Natl. Conf. Adv. Evapotrans., Chicago, IL. 16–17 Dec. 1985. ASAE Publ. no. 14-85. ASAE, St. Joseph, MI.

Reicosky, D.C., and D.B. Peters. 1977. A portable chamber for rapid evapotranspiration measurements on field plots. Agron. J. 69:729–732.

Righes, A.A. 1980. Water uptake and root distribution of soybeans, grain sorghum and corn. M.S. thesis. Iowa State Univ., Ames.

Schoolfield, R.M., P.J.H. Sharp, and C.E. Magnuson. 1981. Non-linear regression of biological temperature-dependent rate models based on absolute reaction-rate theory. J. Theor. Biol. 88:719–731.

Schulze, E.D., N.C. Turner, T. Gollan, and A. Shackel. 1987. Stomatal responses to air humidity and to soil drought. p. 311–321. *In* E. Zeiger et al. (ed.) Stomatal function. Stanford Univ. Press, Stanford, CA.

Sharp, R.E., M.A. Matthews, and J.S. Boyer. 1984. Kok effect and the quantum yield of photosynthesis. Plant Physiol. 75:95–101.

Sivakumar, M.V.K., and S.M. Virmani. 1984. Crop productivity in relation to interception of photosynthetically active radiation. Agric. For. Meteorol. 31:131–141.

Spiertz, J.H.J., and H. van deHaar. 1978. Differences in grain growth, crop photosynthesis and distribution of assimilates between semi-dwarf and a standard cultivar of wheat. Neth. J. Agric. Sci. 26:233–249.

Steiner, J.L. 1986. Dryland grain sorghum water use, light interception and growth responses to planting geometry. Agron. J. 78:720–726.

Unsworth, M.H., V.M. Lesser, and A.S. Heagle. 1984. Radiation interception and the growth of soybeans exposed to ozone in open-top field chambers. J. Appl. Ecol. 21:1059–1079.

Vertregt, N., and F.W.T. Penning de Vries. 1987. A rapid method for determining the efficiency of biosynthesis of plant biomass. J. Theor. Biol. 128:109–119.

Vietor, D.M., R.P. Ariyanayagam, and R.B. Musgrave. 1977. Photosynthetic selection of *Zea mays* L. I. Plant age and leaf position effects and a relationship between leaf and canopy rates. Crop Sci. 17:567–573.

von Caemmerer, S., and G.D. Farquhar. 1981. Some relationships between the biochemistry of photosynthesis and the gas exchange of leaves. Planta 153:367–387.

Walter-Shea, E.A. 1987. Laboratory and field measurements of leaf spectral properties and canopy architecture and their effects on canopy reflectance. Ph.D. diss. Univ. of Nebraska, Lincoln (Diss. Abstr. 87-17268).

Williams, W.A., R.S. Loomis, and C.R. Lepley. 1965. Vegetative growth of corn as affected by population density. I. Productivity in relation to interception of solar radiation. Crop Sci. 5:211–215.

Yao, N.R. 1980. Vegetative and reproductive development of corn at four spring planting dates. M.S. thesis. Pennsylvania State Univ., University Park.

6 Canopy Carbon Assimilation and Crop Radiation-Use Efficiency Dependence on Leaf Nitrogen Content

T. R. Sinclair
USDA-ARS
Gainesville, FL

As with all scientific efforts, the most important step in a modeling exercise is probably an explicit and precise statement of research objectives. My objective in the use of models has been to study quantitatively the possible influence of variation in solar radiation, temperature, and precipitation on crop yield. A number of approaches exist for meeting this objective. Statistical approaches have been used successfully in several studies (e.g., Thompson, 1986). However, difficulties arise because weather variables are confounded and there is no mechanistic basis for studying variations in crop growth processes. Further, statistical models are only reliable for interpolation within data sets based on the mean response. In many cases, questions about crop responses under unusual weather conditions are being asked; extrapolation beyond mean responses is precisely what is desired. On this basis, the use of statistical models can be rejected for projections of crop response to altered climates.

An alternate modeling approach is to use mechanistic models that attempt to simulate processes at a more refined level so that the processes may be integrated to describe the performance of the whole system. Previous chapters have offered good examples of such mechanistic models. However, for describing C input into crop growth models for analysis of climate responses, I have chosen to use a simplified approach. Rather than develop a model that attempts to recreate reality by invoking ever more detailed subprocess models, I have chosen a simple C-assimilation submodel to describe the mechanisms of C input. For my objective of analyzing the influence of weather on crops, Passioura's (1973) admonition to use only a few variables that can all be measured seems especially appropriate.

A simple C-input model may be adequate to predict canopy CO_2-assimilation rates for many studies. Sinclair et al. (1976) compared a complex canopy model, which considered the performance of 300 different

---

 Modeling Crop Photosynthesis—from Biochemistry to Canopy. CSSA Special Publication no. 19.

classes of leaves based on variation in microenvironment, against a model with only two classes of leaves. The simple model segregated leaves between those exposed to direct solar radiation and those under shaded conditions. Calculating the average microenvironment for each of the two classes and combining the CO_2 assimilations resulted in substantial agreement between the simple-model predictions and those of the complex model. In the first section of this chapter, the relationships used to predict canopy CO_2 assimilation and crop biomass accumulation with the simple model will be reviewed.

In many situations, it may be desirable to describe crop biomass accumulation rates without directly recomputing canopy CO_2 assimilation. An approach that is receiving increasing attention is the description of the C input by crops as a radiation-use efficiency (biomass accumulated per unit of solar radiation intercepted). Montieth (1977) suggested that radiation-use efficiency was stable within species, and that there was a fair degree of consistency among species. Consequently, in the second section of this chapter, the methods for estimating crop biomass accumulation will be extended to a derivation of values for radiation-use efficiency for crop canopies. Both theoretical and experimental evidence for summarizing canopy C assimilation as radiation-use efficiency will be reviewed.

Finally, the importance of leaf N content in influencing crop biomass accumulation and radiation-use efficiency will be considered. This is an important issue because leaf CO_2 assimilation rates have been found to be closely linked to leaf N content. The influence of variation in leaf N content can be incorporated directly into the simple canopy CO_2-assimilation model, and the resultant influence on crop biomass accumulation and radiation-use efficiency can be calculated.

SIMPLE CANOPY CARBON DIOXIDE ASSIMILATION MODEL

In this model, developed to estimate daily average values of assimilation rate, the photosynthetic activity of all leaves are calculated by dividing the leaves into only two groups. One group intercepts direct-beam solar radiation and, consequently, is exposed to high irradiances that allow high rates of leaf CO_2 assimilation. The second group of leaves receives only scattered radiation originating from the sky (diffuse radiation) and from other leaves (primarily the leaves intercepting direct-beam radiation). The first group of leaves will be referred to as *sun leaves* and the second group as *shade leaves*. It should be remembered, however, that the actual leaf segments in each group are always changing during the diurnal and seasonal cycles.

The structure of the model proposed to calculate daily, average canopy CO_2-assimilation rates is straightforward. The leaf area of the sun and shade groups is calculated first, followed by an estimation of the average irradiance for each group of leaves. The CO_2-assimilation rate per unit leaf area for each group is calculated from the efficiency of radiation use at low irradiance, the radiation-saturated leaf photosynthetic rate, and average leaf

Table 6–1. BASIC program for a simple C-input model for calculation of mean daily rates of canopy CO_2 assimilation and crop biomass accumulation, and crop radiation-use efficiency (RUE).

```
5   REM CROP BIO
10  REM *** CALCULATION OF CROP BIOMASS ACCUMULATION ***
20  REM *** AND RADIATION-USE EFFICIENCY ***
30  PRINT "INPUT LIGHT-SATURATED LEAF CO2 ASSIMILATION RATE
        (MG CO2 M-2 S-1)"
40  INPUT PS
50  I=.8*SIN(45):REM ASSUMED INCIDENT RADIATION (KJ M-2 S-1)
99  REM
100 REM *** CALCULATE LIGHT INTERCEPTION ***
110 L=4! :REM LAI ASSUMED TO EQUAL 4
120 G=.5 :REM SHADOW PROJECTION = 0.5
130 S=45 :REM WEIGHTED MEAN RADIATION ANGLE
140 F=1-EXP(-L*G/SIN(S)) :REM FRACTION OF INTERCEPTION
150 LSUN=F*SIN(S)/G :REM LAI IN DIRECT RADIATION
160 LSHD=L-LSUN :REM LAI IN SHADE
170 ISUN=I*F/LSUN :REM RADIATION ON SHADE LEAVES
180 ISHD=.2*I*F/LSHD :REM RADIATION ON SHADE LEAVES
199 REM
200 REM *** CALCULATE CO2 ASSIMILATION ***
210 E=5 :REM LIGHT EFFICIENCY (MG CO2 KJ-1)

220 PSUN=PS*(1-EXP(-E*ISUN/PS)) :REM SUN LEAF PHS
230 PSHD=PS*(1-EXP(-E*ISHD/PS)) :REM SHADE LEAF PHS
240 TCO2=PSUN*LSUN+PSHD*LSHD
250 LPRINT "CROP CO2 ASSIMILATION =";TCO2 ;"MG M-2 S-1"
299 REM
300 REM *** CALCULATE BIOMASS AND RUE ***
310 R=.6 :REM ACCOUNT FOR RESPIRATORY LOSSES
320 BM=TCO2*R*30/44
330 LPRINT "BIOMASS PRODUCED = ";BM ; "MG M-2 S-1"
340 RUE=BM/(I*F)
350 LPRINT "RADIATION-USE EFFICIENCY ="; RUE ; "G MJ-1"
400 STOP
PS = 1.0
```

irradiance. Combining the total CO_2 assimilated by each group of leaves gives an estimate of canopy photosynthetic CO_2 assimilation. The final step is to adjust the total canopy CO_2 uptake for respiratory losses of CO_2. Accounting for the CO_2 released by the crop for both maintenance and growth respiration, a respiratory coefficient for a given species is estimated. The net result of these calculations is the biomass accumulation for a particular species.

The BASIC program for this simple canopy CO_2-assimilation model is given in Table 6–1. This program was constructed to provide estimates of daily mean rates of biomass accumulation from an average daily incident solar irradiance. (The program in Table 6–1 can be modified to calculate biomass accumulation rates on an hourly basis, for example, by providing hourly irradiance values [*I*] and calculating the angle of solar elevation [*S*].) Total solar radiation is entered into this version of the model, although photosynthetically active radiation could also be used after numerical adjustments in the model variables. In Table 6–1, the mean daily value of incident solar radiation was calculated from a midday estimate of radiation (0.8 kJ m^{-2} s^{-1}) adjusted by an average sun angle of 45° (Line 50). To calculate the interception of the radiation, the shadow-projection variable, G, was set equal to 0.5 (Line 120) and the sun angle was held at 45° (Line 130). In this example, crop leaf area index was set equal to 4.0 (Line 110).

The segregation of the canopy into sun and shade leaves begins by calculating the fraction of radiation intercepted by the total canopy (Line 140). This equation has been well established as describing the geometry of radiation interception for a canopy with leaves randomly dispersed in the horizontal plane. For leaf canopies where the leaf area index approaches 3 or more, this equation has been shown to depict radiation interception well (e.g., Sinclair & Lemon, 1974; Sinclair & Knoerr, 1982). From basic geometrical considerations, the amount of leaf area that actually intercepts the direct-beam radiation can be calculated (Line 150). By simple difference, the remaining leaf area intercepts only diffuse and scattered radiation (Line 160).

The average irradiances on the sun and shade leaves are calculated by straightforward assumptions concerning the geometry of light distribution. To obtain an average irradiance on sun leaves, it is assumed the intercepted radiation is distributed uniformly over all the sun leaves (Line 170). While this average irradiance is used to calculate CO_2-assimmilation rates of sun leaves, highly precise estimates of irradiance are generally not required because the irradiance is commonly near or at sufficient levels to saturate the CO_2-assimilation rate. Consequently, errors in estimating sun-leaf irradiance do not usually result in large errors in estimated CO_2-assimilation rate. For shade leaves, it is assumed that the total radiation intercepted by the shade leaves is approximately 20% of that intercepted by the sun leaves. This assumption becomes less reliable at the extremes of the proportion of the diffuse component in the incident radiation, especially under overcast conditions when the diffuse component is large. The amount of radiation intercepted by shade leaves is distributed uniformly over all the shade leaves (Line 180). Since the leaf photosynthetic-irradiance response curve is essentially linear at these low irradiances, the assumption of uniform light distribution is usually not important in calculating canopy CO_2-assimilation rates for the shaded leaves.

To calculate CO_2-assimilation rates, an asymptotic exponential equation is used to describe leaf CO_2-assimilation rate for each leaf type in response to incident irradiance (Lines 220 and 230). Boote and Jones (1987) judged this equation to best represent the available leaf CO_2-assimilation data. In addition to the irradiance, the asymptotic exponential equation requires the radiation-use efficiency at low irradiances (E) and radiation-saturated CO_2-assimilation rates (PS). It has been shown for a wide range of species that E is approximately 5.0 g CO_2 MJ^{-1} (Line 210; Ehleringer & Bjorkman, 1977 Ehleringer & Pearcy, 1983). The radiation-saturated rate of CO_2 assimilation is an input in Table 6–1, but it will be shown below that PS is closely linked to leaf N content. Finally, canopy CO_2-assimilation rate is obtained by summing the CO_2-assimilation rates of the sun and shade leaves (Line 240).

The last steps in this model are to calculate the mean daily rate of biomass accumulation by the crop. The losses of CO_2 due to maintenance and growth respiration must be considered. For many crops and cropping situations, the maintenance-respiration rate is about 15 to 20% of the CO_2-assimilation rate of the leaves (e.g., McCree, 1982). The growth-respiration rates are readily defined based on the biochemical composition of the crops (Penning de Vries, 1975). Sinclair and Horie (1989) estimated that growth respiration in soybean [*Glycine max* (L.) Merr.] resulted in 0.65 g of biomass being produced from 1 g of hexose, and in rice (*Oryza sativa* L.) and corn (*Zea mays* L.) about 0.75 g biomass g^{-1} hexose. As a result, the respiratory coefficient for the cereals was about 0.6 (Line 310) and for soybean it was about 0.5. The conversion coefficient for translating CO_2 assimilation to CH_2O is also included in the final calculation of mean daily rate of crop biomass accumulation (Line 320). Finally, the radiation-use efficiency of the crop is calculated by dividing the mean daily rate of crop biomass accumulation by the mean daily solar irradiance (Line 340).

RADIATION-USE EFFICIENCY

Evidence has been accumulating for more than 20 yr that a linear relationship exists between accumulated crop biomass and intercepted solar radiation. Such a linear relatinoship was found for corn in experiments with varying plant populations (Williams et al., 1965) and in soybean for several intervals of crop growth during the season (Shibles & Weber, 1966). Similarly, a linear relationship between accumulated crop biomass and intercepted solar radiation has been demonstrated in wheat (*Triticum aestivum* L.), barley (*Hordeum vulgare* L.), and sugar beet (*Beta vulgaris* L.) (Biscoe & Gallagher, 1977). Monteith (1977) brought attention to this linear relationship in several crops and speculated that for many crops the slope, or radiation-use efficiency (RUE), was approximately 1.4 g MJ^{-1}.

Evaluation of RUE has been done recently in a number of species, showing great stability within a species. Horie and Sakuratani (1985) found RUE of rice to be constant at specific growth stages across 10 growing environments. A study of sorghum [*Sorghum bicolor* (L.) Moench] in three growing environments showed RUE to be constant across environments (Muchow & Coates, 1986). A study of soybean [*Glycine max* (L.) Merr.] vegetative growth showed RUE to be equivalent between 2 yr (Sinclair et al., 1988). Kiniry et al. (1989) calculated RUE for a number of data sets of sunflower (*Helianthus annuus* L.), rice, wheat, corn, and sorghum and concluded RUE was fairly stable within species. With this growing amount of experimental evidence, the utility of describing the potential for C accumulation by crops in terms of RUE has become well established.

The fact that crop biomass accumulation can be expressed simply as a radiation-use efficiency has considerable support from direct observations of canopy photosynthesis. In an early study with rice, Murata (1961) found, for rice crops with full canopies, that a linear relationship existed between canopy CO_2 exchange and solar irradiance. In corn and cotton (*Gossypium hirsutum* L.), a strong correlation between canopy CO_2 assimilation and intercepted radiation has been found across a wide range of irradiances (Hesketh & Baker, 1967). Similarly, Vietor et al. (1977) found corn canopy CO_2-assimilation rates to have a strong linear response to radiation even though there was a quadratic term suppressing continued response at high irradiances. In soybean, Egli et al. (1970) found a highly linear response of canopy photosynthesis to irradiance for three genotypes. A quadratic term in the response function showed suppressed rates at very high irradiances. The more recent study of Baldocchi et al. (1981) with soybean using micrometeorological techniques showed a highly linear response between canopy CO_2-assimilation rates and solar radiation across the entire range of irradiances. In barley, a linear response in canopy CO_2 assimilation and intercepted radiation was found across a wide range of irradiances (Biscoe & Gallagher, 1977). For a tomato [*Lycopersicon lycopersicum* (L.) Karsten] crop, Acock et al. (1978) found an essentially linear response in canopy CO_2 assimilation to incident irradiance. Wilson et al. (1978) found a linear response in subterranean clover (*Trifolium subterraneum* L.) except for an observation at very high irradiances. All observations of canopy CO_2 assimilation at low and moderate irradiances are consistent with a linear response to irradiance and, consequently, with a stable RUE. At very high irradiances, a few observations indicated somewhat less than a linear response to irradiance, but a comparatively limited time of exposure during the diurnal cycle of canopies to these irradiances and the slight deviation in CO_2 assimilation involved results in only a minor non-linear component in daily CO_2 accumulation. Consequently, observed linear responses of CO_2-assimilation rates by canopies to solar irradiance is consistent with a stable RUE.

It is counterintuitive that canopy CO_2-assimilation rates are linearly dependent on irradiance when individual-leaf CO_2-assimilation rates are highly curvilinear and are commonly light saturated at irradiances well below the maximum irradiances that exist in nature. Nevertheless, several geometric analyses for deriving canopy CO_2-assimilation rates from leaf photosynthetic-response curves show essentially linear responses by the canopy to radiation. De Wit (1965) produced the first such analysis and calculated a nearly linear response in canopy CO_2-assimilation rate to the amount of light absorbed. Further, in calculating RUE values from the results presented by de Wit, it is found that the predicted RUE values are stable across the latitudes and growing seasons appropriate for most crops. More recently, Goudriaan (1982) and Horie and Sakuratani (1985), using complex models for summing the photosynthetic activity of individual leaves, derived linear relationships between canopy CO_2-assimilation rates and irradiance. The slopes of these linear relationships are estimates of a stable RUE.

Importantly, the results of de Wit (1965) suggested that RUE varied depending on the radiation-saturated photosynthetic rates of individual leaves. Monteith (1977) summed estimates of leaf CO_2-assimilation rates and showed a dependency of RUE on radiation-saturated leaf photosynthetic rates. The model presented in Table 6–1 can also be used to examine the dependence of RUE on radiation-saturated leaf photosynthetic rates (PS, Lines 30 and 40). Sinclair and Horie (1989) presented the results of such calculations for the cereals, compared with soybean (Fig. 6–1). The curvilinear response of RUE to leaf photosynthesis rates presented in Fig. 6–1 is similar to that presented by Monteith.

The results shown in Fig. 6–1 highlight some important features of crop production resulting from variation in RUE among species. Clearly, the RUE of soybean is less than that of cereals at any given leaf photosynthetic rate because of differences in respiration. In addition, differences in leaf CO_2-assimilation rates among crops will result in variations in RUE. For instance, the radiation-saturated leaf CO_2-assimilation rates of the C_4 species represented by corn are expected to be substantially greater than that of the C_3 species represented by rice. As a consequence, the results in Fig. 6–1 indicate that the RUE of C_4 species would be greater than C_3 species.

A second feature of the curvilinear response in Fig. 6–1 is that, at higher rates of leaf CO_2 assimilation, the value of RUE is fairly insensitive to changes in the leaf assimilation rate. Since nonstressed leaves tend to have high leaf CO_2-assimilation rates, RUE is not expected to be altered greatly by adjustments in leaf assimilation at these high rates. Furthermore, it is anticipated that neither RUE nor crop biomass accumulation can be substantially increased even if very large increases in leaf photosynthetic rates were possible. Fig. 6–1 indicates, for commonly observed leaf photosynthesis rates of about 2.2, 1.4, and 1.4 mg CO_2 m^{-2} s^{-1} for leaves of corn, rice, and soybean, respectively, that the estimated values of RUE are about 1.7 g MJ^{-1} for corn, 1.4 g MJ^{-1} for rice, and 1.2 g MJ^{-1} for soybean.

Figure 6–1 also indicates that decreases in radiation-saturated CO_2-assimilation rates of leaves well below those commonly observed in

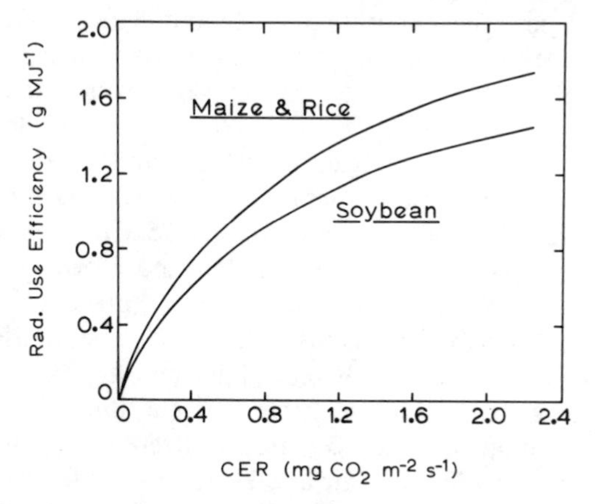

Fig. 6–1. Calculated crop radiation-use efficiency (RUE) on corn, rice, and soybean as a function of radiation-saturated rates of CO_2 exchange rates (CER) for individual leaves (Sinclair & Horie, 1989).

nonstressed crops can result in substantial decreases in RUE. These results indicate that loss of leaf photosynthetic activity due to such factors as leaf maturity and senescence, diseases and pests, or drought stress can cause major decreases in RUE. Consequently, the maintenance of currently achievable leaf CO_2-assimilation rates under field stress conditions appears to be a much more important and fruitful objective than seeking to increase maximum leaf CO_2-assimilation rates.

Importantly for simple models of crop growth, Fig. 6–1 provides a very useful function for calculation of the C input to a crop. Either RUE can be assumed to be stable when considering a nonstressed crop (e.g., Spaeth et al., 1987), or input information for the stress response function for radiation-saturated rates of leaf CO_2 assimilation can be used to predict RUE (e.g., Sinclair, 1986). Crop growth is then readily calculated from the value of RUE and estimates of daily amounts of intercepted solar radiation.

LEAF NITROGEN AND CARBON ASSIMILATION

It was shown above that crop biomass accumulation expressed as RUE is directly dependent on radiation-saturated rates of leaf photosynthesis. Here, the dependence of leaf CO_2-assimilation rates on leaf N content will be reviewed. It is well recognized that the quantity of photosynthetic apparatus in a leaf correlates well with the radiation-saturated rates of leaf CO_2 assimilation. Many studies have shown a close correlation between leaf CO_2-assimilation rates and ribulose-1,5-bisphosphate carboxylase activity (e.g., Massacci et al., 1986; Ford & Shibles, 1988). Since ribulose-1,5-bisphosphate carboxylase alone may account for up to 50% of the soluble protein in leaves (Schmitt & Edwards, 1981), a good correlation between leaf CO_2-assimilation rate and leaf protein or N content is expected.

Many studies show a good correlation between leaf CO_2-assimilation rate and N content per unit leaf mass (e.g., Ojima et al., 1965; Boote et al., 1978). For consistency in units, however, it is necessary to express N content per unit leaf area. In reviewing the literature in which the relationship between radiation-saturated leaf CO_2-assimilation rate and leaf N content for soybean and rice was investigated, Sinclair and Horie (1989) found a consistently close correlation. Generally, the correlation coefficients between the two variables were greater than 0.75. In comparing various experimental treatments and cultivars, Sinclair and Horie concluded that, within a species, a single response function well represented many of the observations (Fig. 6–2). In each case, leaf CO_2-assimilation rate was dependent approximately linearly on leaf N per unit area at lower N levels. At higher N contents, the response curve was curvilinear, approaching a maximum CO_2-assimilation rate. Distinctly different response functions existed among species, which Sinclair and Horie speculated were a result of differences in photosynthetic pathway and leaf structure.

If all leaves in the crop canopy contain the same N content, then it is possible to predict RUE directly from leaf N content by means of the

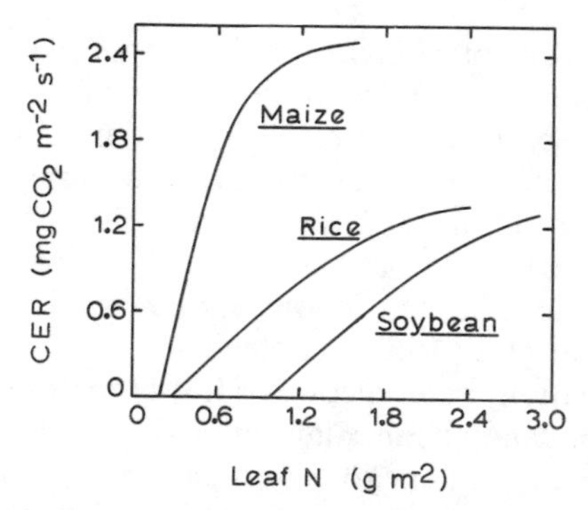

Fig. 6–2. Carbon dioxide exchange rates (CER) of individual leaves of corn, rice, and soybean as a function of individual leaf N contents per unit leaf area (Sinclair & Horie, 1989).

response function for radiation-saturated rates of leaf CO_2 assimilation. The single modification required in the model presented in Table 6–1 is the inclusion of an equation as represented in Fig. 6–2 (Sinclair & Horie, 1989) giving the dependency of radiation-saturated leaf photosynthesis rate (PS) on leaf N content per unit leaf area for each species. The BASIC program in Table 6–1 is modified by having leaf N be the input value (Line 40) and adding a new line to calculate PS as a function of leaf N content.

Sinclair and Horie (1989) calculated the dependency of RUE on leaf N content using this approach, and a curvilinear response between the two variables was predicted (Fig. 6–3). At lower leaf N contents, RUE was essentially linearly dependent in leaf N. Muchow and Davis (1988) found, for both corn and sorghum in a fertilizer experiment, a high linear correlation between RUE and the N content per unit leaf area. On the other hand, Fig. 6–3 indicates that, at high leaf N contents, RUE approaches a maximum value. Importantly, Fig. 6–3 illustrates the dramatic changes in RUE that

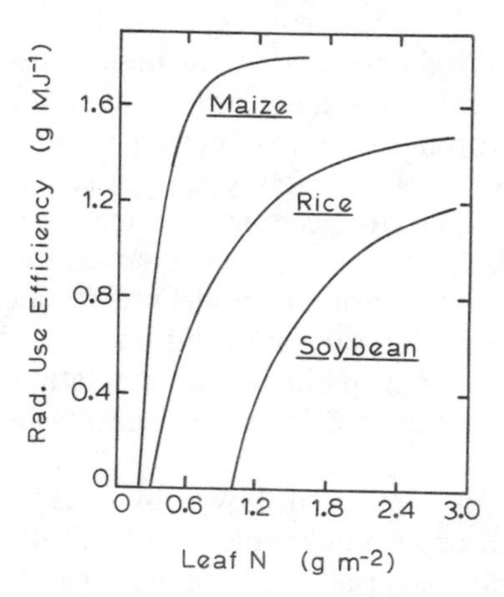

Fig. 6–3. Estimated crop radiation-use efficiency (RUE) for corn, rice, and soybean as a function of individual leaf N contents per unit leaf area (Sinclair & Horie, 1989).

can happen as leaf N contents become less than those giving maximum RUE. Decreasing leaf N content from such stresses as inadequate soil fertility, diseases, or leaf senescence should result in substantial decreases in RUE.

Important differences among species are also clearly illustrated in Fig. 6-3. The RUE value of corn was highest at any given leaf N content. Rice had greater RUE values than soybean at any leaf N. These differences indicate that variation in N-use efficiency among species will result because of the differences in RUE at given leaf N contents.

The RUE calculations as a function of leaf N content assumed uniform leaf N throughout the canopy. Since it is unlikely that such a situation exists, detailed analyses have been done to determine the influence of various leaf N distributions in the canopy on crop CO_2 assimilation. Field (1983) showed optimization of leaf N distribution for maximum canopy CO_2 assimilation occurs when the highest leaf N content is at the top of the canopy with progressively lower contents down to the bottom of the canopy. Field concluded, however, that the optimized leaf N distribution resulted in only a 3% greater daily C gain than a uniform leaf N content throughout the canopy. The analysis of photosynthesis distribution in a canopy by Gutschick and Wiegel (1988) similarly indicated a 5% increase in canopy CO_2 assimilation with an optimized distribution in leaf photosynthetic activity, compared with a uniform distribution in activity. Hirose and Werger (1987) studied the optimum N distribution using a light-attenuation model that gave uniform irradiances over all leaves at a particular depth in the canopy. They showed that the relative advantage in optimization of the leaf N distribution increased with leaf area index. Overall, the analyses done thus far indicate only very small advantages in optimizing the vertical distribution of leaf N, compared with a uniform N-content distribution.

SUMMARY

To a reasonable approximation, it seems possible to calculate canopy CO_2-assimilation rates on a daily basis using simple models. The simple model, segregating the crop canopy into either sun or shade leaves, was used to calculate daily mean rates of canopy CO_2 assimilation. Simplifying assumptions were also used to account for maintenance and growth respiration, so that daily rates of crop biomass accumulation were calculated. The final step of this simple model was to calculate the RUE of a crop based on either radiation-saturated leaf photosynthesis rates or leaf N content. Summarization of crop biomass accumulation by RUE provides a very useful tool for estimating C input in studies using models of crop growth.

Radiation-use efficiency has been shown, by several approaches, to be a satisfactory method for describing crop C input. A substantial amount of data directly supports the view that the biomass accumulated by a crop is linearly dependent on the amount of intercepted solar radiation. The value of RUE, in fact, seems to be fairly stable within a species across a range of conditions for nonstressed crops. Several types of theoretical analyses for summarizing leaf photosynthesis rates in a cropy canopy support the conclusion of stability in RUE. Even though these analyses show a dependency of RUE on radiation-saturated rates of leaf CO_2 assimilation, in the usual range of leaf photosynthesis rates observed for nonstressed crops little variation in RUE is expected (Fig. 6-1). These results are discouraging for hypothesized crop yield increases resulting from increased leaf photosynthesis rates. On the other hand, decreasing radiation-saturated rates of leaf CO_2 assimilation can have important negative effects on RUE.

Importantly for attempts to use a simple C-input submodel for studying crop growth, it seems that the simulation requirements can be limited to a rather straightforward response function for RUE (dependent on leaf N and stresses), fraction of intercepted radiation (dependent on leaf area), and incident solar radiation. Since both RUE and leaf area growth are dependent on N availability and allocation in the plant, the complexity of the model may need to be increased to depict the various N processes in the soil and plants. Nevertheless, as a simple approach for modeling C input, RUE provides a very useful and powerful tool for doing mechanistic analyses of crop response to climatic variation.

REFERENCES

Acock, B., D.A. Charles-Edwards, D.J. Fitter, D.W. Hand, L.S. Ludwig, J.W. Wilson, and A.C. Withers. 1978. The contribution of leaves from different levels within a tomato crop to canopy net photosynthesis: An experimental examination of two canopy models. J. Exp. Bot. 29:815–827.

Baldocchi, D.D., S.B. Verma, and N.J. Rosenberg. 1981. Mass and energy exchanges of a soybean canopy under various environmental regimes. Agron. J. 73:706–710.

Biscoe, P.V., and J.N. Gallagher. 1977. Weather, dry matter production, and yield. p. 75–100. *In* J.J. Landsberg and C.V. Cutting (ed.) Environmental effects on crop physiology. Academic Press, New York.

Boote, K.J., R.N. Gallaher, W.K. Robertson, K. Hinson, and L.C. Hammond. 1978. Effect of foliar fertilization on photosynthesis, leaf nutrition, and yield of soybeans. Agron. J. 70:787–791.

Boote, K.J., and J.W. Jones. 1987. Equations to define canopy photosynthesis from quantum efficiency, maximum leaf rate, light extinction, leaf area index, and photon flux density. p. 415–418. *In* J. Biggins (ed.) Progress in photosynthetic research. Vol. 4. Martinus Nijhoff Publ., The Hague.

deWit, C.T. 1965. Photosynthesis of leaf canopies. Agr. Res. Rep. no. 663. PUDOC, Wageningen, the Netherlands.

Egli, D.B., J.W. Pendleton, and D.B. Peters. 1970. Photosynthetic rate of three soybean communities as related to carbon dioxide levels and solar radiation. Agron. J. 62:411–414.

Ehleringer, J., and O. Bjorkman. 1977. Quantum yields in C_3 and C_4 plants. Dependence on temperature, CO_2, and O_2 concentration. Plant Physiol. 59:86–90.

Ehleringer, J., and R.W. Pearcy. 1983. Variation in quantum yield for CO_2 uptake among C_3 and C_4 plants. Plant Physiol. 73:555–559.

Field, C. 1983. Allocating leaf nitrogen for the maximization of carbon gain: Leaf age as a control on the allocation program. Oecologia 56:341–347.

Ford, D.M., and R. Shibles. 1988. Photosynthesis and other traits in relation to chloroplast number during leaf senescence. Plant Physiol. 86:108–111.

Goudriaan, J. 1982. Potential production processes. p. 98–113. *In* F.W.T. Penning de Vries and H.H. van Laar (ed.) Simulation of plant growth and crop production. PUDOC, Wageningen, the Netherlands.

Gutschick, V.P., and F.W. Wiegel. 1988. Optimizing the canopy photosynthetic rate by patterns of investments in specific leaf mass. Am. Nat. 132:67–86.

Hesketh, J., and D. Baker. 1967. Light and carbon assimilation by plant communities. Crop Sci. 7:285–293.

Hirose, T., and M.J.A. Werger. 1987. Maximizing daily canopy photosynthesis with respect to the leaf nitrogen allocation pattern in the canopy. Oecologia. 72:520–526.

Horie, T., and T. Sakuratani. 1985. Studies on crop–weather relationship model in rice. (1) Relation between absorbed solar radiation by the crop and the dry matter production. Jpn. J. Agric. Meterol. 40:331–342.

Kiniry, J.R., C.A. Jones, J.C. O'Toole, R. Blanchet, M. Cabelguenne, and D.A. Spanel. 1989. Radiation use efficiency in biomass accumulation prior to grain filling for five grain-crop species. Field Crops Res. 20:51–64.

Massacci, A., M.T. Giardi, D. Tricoli, and G. DiMarco. 1986. Net photosynthesis, carbon dioxide compensation point, dark respiration, and ribulose-1,5-bisphosphate carboxylase activity in wheat. Crop Sci. 26:557–563.

McCree, K.J. 1982. Maintenance requirement of white clover at high and low growth rates. Crop Sci. 22:345–351.

Monteith, J.L. 1977. Climate and the efficiency of crop production in Britain. Philos. Trans. R. Soc. London B 281:277–294.

Muchow, R.C., and D.B. Coates. 1986. An analysis of the environmental limitation to yield of irrigated grain sorghum during the dry season in tropical Australia using a radiation interception model. Aust. J. Agric. Res. 37:135–148.

Muchow, R.C., and R. Davis. 1988. Effect of nitrogen supply on the comparative productivity of maize and sorghum in a semi-arid tropical environment. II. Radiation interception and biomass accumulation. Field Crops Res. 18:17–30.

Murata, Y. 1961. Studies on the photosynthesis of rice plants and its culture significance. Bull. Nat. Inst. Agric. Sci. Ser. D, Physiol. Genet. 9:1–170.

Ojima, M., J. Fukui, and I. Watanabe. 1965. Studies on the seed production in soybean. II. Effect of three major nutrient elements supply and leaf age on the photosynthetic activity and diurnal changes in photosynthesis of soybeans under constant temperature and light intensity. Proc. Crop Sci. Soc. Jpn. 33:437–442.

Passioura, J.B. 1973. Sense and nonsense in crop simulation. J. Aust. Inst. Agric. Sci. 39:181–183.

Penning de Vries, F.W.T. 1975. Use of assimilates in higher plants. p. 459–480. *In* J.P. Cooper (ed.) Photosynthesis and productivity in different environments. Cambridge Univ. Press, Cambridge, England.

Schmitt, M.R., and G.E. Edwards. 1981. Photosynthetic capacity and nitrogen use efficiency of maize, wheat, and rice: A comparison between C_3 and C_4 photosynthesis. J. Exp. Bot. 32:459–466.

Shibles, R.M., and C.R. Weber. 1965. Leaf area, solar radiation interception, and dry matter production by soybeans. Crop Sci. 5:575–578.

Sinclair, T.R. 1986. Water and nitrogen limitations in soybean grain production. I. Model development. Field Crops Res. 15:125–141.

Sinclair, T.R., and T. Horie. 1989. Leaf nitrogen, photosynthesis, and crop radiation use efficiency: A review. Crop Sci. 29:90–98.

Sinclair, T.R., and K.R. Knoerr. 1982. Distribution of photosynthetically active radiation in the canopy of a loblolly pine plantation. J. Appl. Ecol. 19:183–191.

Sinclair, T.R., and E.R. Lemon. 1974. Penetration of photosynthetically active radiation in corn canopies. Agron. J. 66:201–205.

Sinclair, T.R., R.C. Muchow, J.M. Bennett, and L.C. Hammond. 1988. Relative sensitivity of nitrogen and biomass accumulation to drought in field-grown soybean. Agron. J. 79:986–991.

Sinclair, T.R., C.E. Murphy, Jr., and K.R. Knoerr. 1976. Development and evaluation of simplified models for simulating canopy photosynthesis and transpiration. J. Appl. Ecol. 13:813–839.

Spaeth, S.C., T.R. Sinclair, T. Ohnuma, and S. Konno. 1987. Temperature, radiation, and duration dependence of high soybean yields: Measurement and simulation. Field Crops Res. 16:297–307.

Thompson, L.M. 1986. Climatic change, weather variability, and corn production. Agron. J. 78:649–653.

Vietor, D.M., R.P. Ariyanayagam, and R.B. Musgrave. 1977. Photosynthetic selection of *Zea mays* L. I. Plant age and leaf position effects and a relationship between leaf and canopy rates. Crop Sci. 17:567–573.

Williams, W.A., R.S. Loomis, and C.R. Lepley. 1965. Vegetative growth of corn as affected by population density. I. Productivity in relation to interception of solar radiation. Crop Sci. 5:211–215.

Wilson, D.R., C.J. Fernandez, and K.J. McCree. 1978. CO_2 exchange of subterranean clover in variable light environments. Crop Sci. 18:19–22.

7 The Prediction of Canopy Assimilation

K. J. Boote

University of Florida
Gainesville, FL

R. S. Loomis

University of California
Davis, CA

There is an increased interest in predicting the growth of crops and natural ecosystems in response to climatic and soil-related factors. Integrative simulation models have been central and essential tools for integrating information across levels of organization and making quantitative predictions. The essential core of most vegetation models is the system's use and balance of C, beginning with the input of C from canopy assimilation.

In this publication, the contributing authors have summarized a number of approaches used to predict leaf and canopy photosynthesis. Most of their models can stand alone for studies of photosynthesis, or they can be incorporated into crop models. In this chapter, we will identify and compare the diversity of approaches used by the respective authors. We will review the processes involved, the environmental inputs considered, and the characteristics of leaves and crop canopies as they relate in a systems approach. Where similar approaches are used, we view this as a consensus of opinion. Where the authors' approaches differ, we will attempt to gain insight into the uncertainties or lack of knowledge that resulted in multiple approaches.

A Systems Approach: Processes and State Variables

Processes: The photosynthetic processes include light penetration through leaves and canopy, photon capture, photon excitation of chlorophyll and resultant charge separation, electron transport, use of energy from electron transfer to synthesize adenosine triphosphate (ATP) and nicotinamide adenine dinucleotide phosphate (NADPH), activation of various enzymes via light-produced metabolites, CO_2 diffusion from the gas phase to the chloroplast,

 Modeling Crop Photosynthesis—from Biochemistry to Canopy. CSSA Special Publication no. 19.

enzyme processing of CO_2 or O_2, and enzymatic steps in the reduction and interconversion of C compounds.

State Variables: Prediction of assimilation is dependent on the current state of the photosynthetic system. The state variables include leaf characteristics (amounts of photon-absorbing pigments, electron-transport intermediates, and ribulose-1,5-bisphosphate [RuBP] carboxylase-oxygenase [rubisco] protein per unit leaf area) and canopy characteristics (amount of leaf area per unit land area, leaf angle, canopy dimensions). Leaf and canopy characteristics change over time and are themselves influenced by the environmental conditions (irradiance, CO_2 concentration, and temperature) occurring during plant growth.

Inputs: Environmental inputs that drive photosynthetic assimilation at the leaf or canopy level include photosynthetic photon flux density (PPFD), atmospheric concentrations of CO_2 and O_2, and temperature.

Time Scale: Models of leaf and canopy assimilation are most commonly quasi-steady-state models, being driven by light flux and by current levels of other inputs. Many crop models, by contrast, have been constructed with daily time steps, partly for simplicity and partly because of the daily nature of Class-A weather-station inputs available. Three chapters in this book, (Ch. 4, Gutschick; Ch. 5, Norman & Arkebauer; and Ch. 6, Sinclair) attempt to derive daily assimilation from instantaneous leaf-level parameters. Inclusion of highly detailed canopy models within the crop model is now quite practical, although it increases computer run times (Denison & Loomis, 1989; Norman & Arkebauer, Ch. 5 in this book). Alternatives include reducing the output of the detailed model to look-up tables (Ng & Loomis, 1984) or using mathematical techniques to integrate (Gutschick, Ch. 4 in this book). For some purposes, very simple models can be used at the cost of reducing the detail of processes (Sinclair, Ch. 6 in this book). Sinclair's approach depends only on radiation, bypassing the detail of rubisco kinetic response to CO_2 and O_2 concentrations, and temperature. In that way, he achieved a simple, summary model of production.

LEAF-LEVEL PREDICTION

Light Response of Single Leaves

In leaves, the absorption of photons by pigments in the thylakoid membranes is coupled to electron transfer across the thylakoid membranes, leading to production of NADPH. Coupled to the electron transfer is the transfer of protons into the intrathylakoid channels. This proton accumulation subsequently drives the synthesis of ATP. These two high-energy nucleotides, NADPH and ATP, are used in the C fixation and reduction cycle to regenerate the CO_2 acceptor, RuBP, and to reduce 3-phosphoglyceric acid. The perennial question, partially addressed by

Farquhar et al. (1980), is to understand when assimilation rate is limited by lack of light, or by lack of CO_2, or colimited by both. Taylor and Terry (1984) pointed out that assimilation response is frequently colimited by both light and CO_2, and that PPFD has an effect on slope of response to CO_2 concentration even in the CO_2-limiting region. They concluded that photochemical energy supply plays a role even in the CO_2-limiting region via the activation of rubisco enzyme. Additionally, CO_2 assimilation is affected by temperature and O_2 concentration.

Leaf light-response curves are generally accepted to have: (1) a light-limited region in which light-utilization efficiency (quantum efficiency = Q_E) is greatest and almost constant, and (ii) a light-saturated region (P_{max}) in which further increases in light fail to increase photosynthesis (P). The difficulty here is that both Q_E and P_{max} are affected by temperature, and CO_2 and O_2 concentrations. Furthermore, P_{max} is affected by light and temperature history through their influences on leaf thickness, N content, and other state variables of the leaf system.

Predicting electron transport and the maximum light-saturated rate of electron transport (J_{max}) represents a more basic approach than predicting P_{max} (in terms of CO_2 fixation) because energy from electron-transport processes is used to drive both carboxylation and oxygenation by rubisco. Gerbaud and Andre (1979, 1980) concluded that, at a given irradiance, O_2 and CO_2 compete for reducing power produced at a constant rate by the light reactions. The sum of CO_2 fixed and O_2 fixed was nearly equal to the gross O_2 evolution (measured by $^{18}O_2$ isotope) even as CO_2 concentration was increased from low to high. In order to bypass the problem of interacting effects of CO_2 and O_2 concentrations on the P_{max} and Q_E, several authors in this volume (Evans and Farquhar, Ch. 1; Harley and Tenhunen, Ch. 2) define equations to predict electron transport in response to PPFD. At low irradiance, the initial increase in electron transport is linear, because there are few multiple photon hits per individual photosystem; thus, photosystems and associated electron-transfer carriers are generally more available to process photons or electrons, respectively. This could be defined as a type of ''electron-transfer'' quantum efficiency. The term J_{max} represents the maximum rate of electron transport when all light-harvesting complexes, electron-transport carriers, and CO_2-processing enzymes are saturated.

Equations for Predicting Leaf Photosynthesis Response to Photosynthetic Photon Flux Density

Three of our authors (Evans & Farquhar, Ch. 1; Norman & Arkebauer, Ch. 5; Gutschick, Ch. 4) use nonrectangular-hyperbola equations similar to that shown below, that include the parameters P_{max}, Q_E, and a curvature factor Θ. The nonrectangular hyperbola is the lower root of the quadratic equation:

$$\Theta P^2 - (Q_E \text{ PPFD} + P_{max}) P + Q_E \text{ PPFD } P_{max} = 0 \qquad [1]$$

which is

$$P = \{Q_E \text{ PPFD} + P_{max} - [(Q_E \text{ PPFD} + P_{max})^2 - 4\,\Theta\, Q_E \text{ PPFD}\, P_{max}]^{1/2}\}/(2\,\Theta) \quad [2]$$

Johnson and Thornley (1984) reviewed the use of the nonrectangular hyperbola for predicting photosynthetic response to PPFD. When Θ is zero, the equation becomes a rectangular hyperbola (the familiar Michaelis-Menten equation). With $\Theta = 1.0$, the equation is a Blackman response of two intersecting straight lines. Evans and Farquhar (Ch. 1 in this book) use the nonrectangular hyperbola to describe electron transport (J) as a function of incident PPFD, J_{max}, and $\Theta = 0.7$. Gutschick, for his canopy assimilation model, used $\Theta = 0.9$ for leaf response to light. Most evidence suggests that the asymptotic exponential or a nonrectangular hyperbola with $\Theta = 0.7$ to 0.9 provides a better fit than the Michaelis-Menten equation (Marshall & Biscoe, 1980; Peat, 1970). In Fig. 7-1, photosynthetic response to PPFD is illustrated using the nonrectangular hyperbola ($\Theta = 0.0$, 0.7, 0.9, and 1.0) and the asymptotic exponential equation

$$P = P_{max}\,[1 - \exp(-\,Q_E \text{ PPFD}/P_{max})] \quad [3]$$

The asymptotic exponential is similar to nonrectangular hyperbolae having curvatures between 0.7 and 0.9.

In the nonrectangular hyperbola initially proposed by Rabinowitch (1951), Θ represents the ratio of physical resistance to CO_2 diffusion (r_d) to the physical plus carboxylation resistance (r_x):

$$\Theta = r_d/(r_d + r_x) \quad [4]$$

In this equation, a curvature near 1.0 implies very little carboxylation resistance, whereas a value near zero implies very small physical resistance.

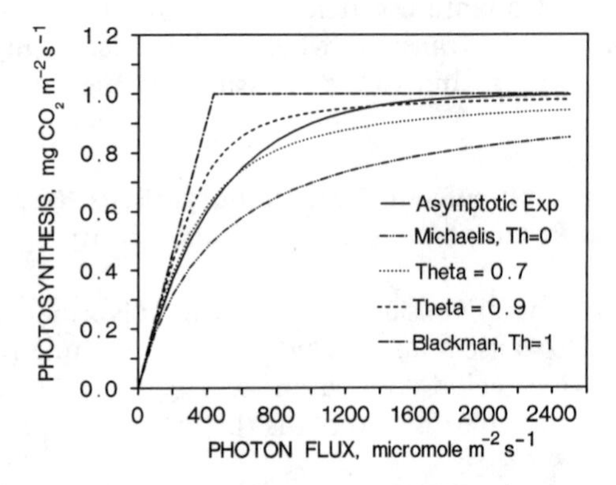

Fig. 7-1. Comparison of leaf-assimilation response to PPFD using asymptotic exponential and nonrectangular-hyperbola equations with curvatures of 0 (Michaelis-Menten), 0.7, 0.9, and 1.0 (Blackman).

In practice, stomatal sensitivity to intercellular CO_2 concentration (C_i) probably serves as a moderating feedback loop so that the value for Θ is hardly ever close to zero, but is rather closer to the values of 0.7 to 0.9 used by the authors in this volume. In either case, the curvature value is obtained from experimental measurements and may vary from case to case.

In practice, J_{max} and the curvature for electron transport vs. irradiance are determined from leaf photosynthetic response to irradiance when measured under saturating CO_2 concentration. This allows maximum electron-transport capacity to be expressed under conditions that minimize the use of electrons for oxygenation. Harley and Tenhunen (Ch. 2 of this book) do not solve for electron transfer per se, but solve for P_{ml}, the light- and CO_2-saturated assimilation rate, and for α, the initial slope of the curve of CO_2-saturated assimilation vs. incident PPFD. They also have demonstrated how to derive these parameters from experimental data (Harley et al., 1985; Tenhunen et al., 1976). Hopefully, future research will reveal how we can predict P_{ml} or J_{max} of a leaf from its physical and biochemical composition and as a function of growth environment, rather than obtaining them each time from measured leaf assimilation curves.

Relationship of Maximum Electron Transport and Photosynthesis to Leaf Physical and Biochemical Composition

Evans and Farquhar (Ch. 1 of this book) suggest that J_{max}, like P_{max}, is a property of thylakoids that depends on growth conditions. They propose that J_{max} is proportional to the chlorophyll and N content per unit leaf area because those parameters reflect the amount of photosynthetic apparatus per unit leaf area. Within a species, the amount of rubisco per unit leaf area is also proportional to the amount of leaf N per unit leaf area. Using similar assumptions, Sinclair (Ch. 6 of this book) assumed P_{max} to be a function of leaf N per unit leaf area, and predicted canopy assimilation response to leaf N. Evans and Farquhar (Ch. 1 of this book) proposed that growth-irradiance history has additional effects on J_{max} beyond its effect on the amount of chlorophyll, N, and rubisco per unit leaf area. Thylakoid membranes of a leaf grown under low irradiance are enriched in light-harvesting chlorophyll a/b protein complex, and depleted in Photosystem II reaction center complexes and electron-transfer components that lead to synthesis of ATP and NADPH (Anderson, 1986). Acock (Ch. 3 of this book) allowed leaf conductance to CO_2 to be a function of the light history of the leaf tissue.

Quantum Efficiency

There is fairly common agreement that Q_E (on an absorbed-PPFD basis) is nearly constant among C_3 species when determined under standard CO_2 and O_2 concentration conditions and standard temperature. The value of Q_E is 0.052 mol mol^{-1} at 330 μL L^{-1} CO_2 concentration, 21% O_2 concentration, and 30 °C (Ehleringer & Björkman, 1977; Ehleringer & Pearcy,

1983; Björkman & Demmig, 1987; Evans, 1987). Evans (1987) used a correction of 0.85 because Q_E in the photosynthetically active fraction of sunlight is 15% less than Q_E measured in 600-nm irradiance. The α value of Harley and Tenhunen is similar and proportional to Q_E, except on an incident-light basis. Most of the authors in this volume accepted the idea of constancy of Q_E (at saturating CO_2 concentration) and focused on the problem of predicting the rate of J_{max} or of CO_2 assimilation at its maximum rate (P_{max}) as described above. The CO_2-concentration dependence of Q_E in 21% O_2 follows Michaelis-Menten kinetics similar to that of rubisco enzyme; this CO_2-concentration dependence has been described by Ehleringer and Björkman (1977), Ehleringer and Pearcy (1983), Peisker et al. (1983), and Kirschbaum and Farquhar (1987). Temperature effects on Q_E at standard CO_2 and O_2 concentration are described by Ehleringer and Björkman (1977) and Ehleringer and Pearcy (1983). Temperature and CO_2 and O_2 concentrations would, therefore, be expected to influence the initial slopes (Q_E), curvatures (Θ), and maxima (P_{max}) for the light-response curves in Fig. 7-1.

Leaf Response to Carbon Dioxide and Oxygen Concentrations

All of the authors in this volume except two, Acock (Ch. 3) and Sinclair (Ch. 6), used the Farquhar approach to predict leaf photosynthetic response to C_i. Acock predicts gross photosynthesis and photorespiration as a function of external CO_2 concentration, and in that way accounts for the effects of both CO_2 and O_2 concentrations. Acock's Eq. [21], [22], [23], and [24] create much the same sensitivity of photosynthesis to CO_2 and O_2 concentrations, and to temperature, as do Eq. [2], [3], [4], and [5] of Harley and Tenhunen (Ch. 2). Sinclair's simple model is based on leaf and canopy assimilation at ambient CO_2 concentration (i.e., with no variation in CO_2 concentration).

Evans and Farquhar (Ch. 1 of this book) and Farquhar and von Caemmerer (1982) assume that net leaf CO_2-assimilation rate (A) essentially falls into two regions, one in which A responds almost linearly to increasing C_i (here A is limited to rubisco activity), and a second region in which A responds less rapidly to increasing C_i, because the rate is limited by electron transport (J) and the regeneration of RuBP. When electron transport is limiting, they use the following equation:

$$A = J\,(C_i - \Gamma^*)/(4.5C_i + 10.5\Gamma^*) - R_d \qquad [5]$$

where Γ^* is the CO_2 compensation point in the absence of mitochondrial respiration. The derivation of the values of 4.5 and 10.5 were explained by von Caemmerer and Farquhar (1981). These values are based on the electron-transport rate to supply ATP and NADPH consumption by carboxylation and photorespiration, assuming that one ATP is produced per three H^+ generated. In the absence of O_2, a minimum of four electrons are required (for reductant) per CO_2 fixed; this minimum increases to 4.5

electrons to provide three ATP per CO_2 fixed if only one ATP is produced per three H^+. The Γ^* is influenced by O_2 and temperature, and is the point at which two oxygenations occur per CO_2 fixed. Depending on O_2 concentration, some of the energy of electron transport will drive O_2 fixation and photorespiration and increase total electron requirement. The value 10.5 reflects this 2:1 ratio of oxygenation to CO_2 fixation at Γ^*, plus the use of additional ATP energy in the photorespiration cycle.

Electron-transport rate can also be estimated experimentally by chlorophyll fluorescence techniques (Sharkey et al., 1988) or gross O_2 evolution techniques (Gerbaud and Andre, 1980). Sharkey et al. (1988) found that electron transport estimated by fluorescence was directly comparable (nearly 1:1) with electron transport estimated by CO_2 assimilation (J_c) using the following equation also presented by von Caemmerer and Farquhar (1981):

$$J_c = 4\,(A + R_d)\,(C_i + 2\Gamma^*)/(C_i - \Gamma^*) \qquad [6]$$

This equation assumes only four electrons are used per CO_2 fixed (in the absence of O_2). Except for assumptions regarding electrons per CO_2 fixed (and ATP/H^+), Eq. [5] and [6] are comparable. Sharkey et al. (1988) showed that the electron-transport rate was reduced as C_i was reduced below 300 $\mu L\ L^{-1}$, but only if the PPFD was fairly high (> 380 $\mu mol\ m^{-2}\ s^{-1}$). If light was low, C_i had no effect on electron transport. Electron-transfer rate was greater in 21% O_2 than in 2% O_2, but only if PPFD was fairly high. Using $^{18}O_2$ techniques, Gerbaud and Andre (1980) showed that gross O_2 evolution (an estimate of electron-transport rate) was reduced when external CO_2 concentration fell below 330 $\mu L\ L^{-1}$ (if in 2% O_2) or below 100 $\mu L\ L^{-1}$ (if in 21% O_2). They proposed that rubisco was inactivated at low CO_2 concentrations, but we suspect electron transport was more likely decreased because O_2 or CO_2 were insufficiently available to allow maximum electron transfer.

When rubisco rather than electron transfer is limited, Evans and Farquhar (Ch. 1 of this book) assume the dependence of A on C_i to be given by

$$A = V_{max}\,(C_i - \Gamma^*)/[C_i + K_c(1 + O/K_o)] - R_d \qquad [7]$$

where V_{max} is the rubisco activity with saturating CO_2 concentration, K_c and K_o are the Michaelis constants for CO_2 and O_2, respectively, and O is the O_2 concentration (intercellular O_2 is assumed the same as ambient). Evans and Farquhar also assume that assimilation will be the minimum of either Eq. [5] or [7], whichever is more limiting. This approach gives a somewhat abrupt transition from the rubisco-limiting region to the electron-transport-limiting region, because of the simplifying assumption of a linear increase in rubisco velocity with increasing RuBP concentration until RuBP supply matches rubisco sites. As will be discussed in the next section, the fraction of active sites is influenced by light energy and ATP supply. This transition often occurs very near the level of ambient CO_2 concentration and results in a slight misfit there, although the model fit is reasonable if both parts of the equation are separately solved from the appropriate data.

Harley and Tenhunen (Ch. 2 of this book) also adopted the Farquhar and von Caemmerer (1982) approach, with minor modifications and one additional limitation to assimilation, based on limited availability of orthophosphate (P_i) for phosphorylation and regeneration of RuBP. The situation of unavailable P_i occurs if the P_i is sequestered as sugar phosphates (Sharkey, 1985). The latter occurs if the plant is unable to synthesize sucrose (for export) and starch as fast as CO_2 is assimilated. This situation may occur if growth is limited by nonoptimum temperature, if sink tissues have been removed, or if CO_2 concentration is increased. Sharkey suggested that P_i deficit does not cause RuBP shortage; rather, ATP becomes deficient and the activation state of rubisco declines as CO_2 concentration is increased. The photosynthetic limitation from failure to rapidly recycle triose phosphate and P_i can be demonstrated experimentally (photosynthesis becomes insensitive to the typical O_2 inhibition). The problem from a crop modeling viewpoint is how to predict its occurrence a priori, without having to do the measurement to prove it.

Colimitations of Light and Carbon Dioxide Concentration and Transition from Carbon Dioxide-Limited to Light-Limited Regions

The model of Farquhar et al. (1980) has served as a good working hypothesis and simplification that has stimulated much useful research on RuBP levels, metabolite levels, and rubisco-activation state as a function of CO_2 concentration and light. As noted above, their approach gives an A/C_i curve with a somewhat abrupt transition between the CO_2-limiting region and the electron-transport-limiting region. Activation of rubisco (by CO_2 and light) and depletion of the light-activating energy supply as CO_2 assimilation increases are features that affect the transition between the two regions. A primary key to this transition is that light activation of rubisco by rubisco activase (a soluble protein) is dependent on ATP supply (Streusand & Portis, 1987). For example, the activation state of rubisco is minimal in low light and increases to an asymptote as light increases to about 1500 μmol photons m^{-2} s^{-1} (Perchorowicz & Jensen, 1983; Taylor & Terry, 1984). Under low light where RuBP concentration is assumed to be limiting in the Farquhar model, rubisco activity is very much decreased (Perchorowicz & Jensen, 1983) and RuBP concentration remains higher than predicted. The net effect is that light activation of rubisco controls the carboxylation rate so that RuBP concentration remains relatively constant under varying light levels, and so that the slope of the A/C_i curve varies with light flux even in the hypothetical "CO_2-limited" region. Several researchers (Taylor & Terry, 1984; Weber et al., 1987) have reported that the initial slope of the A/C_i curve is strongly dependent on irradiance (consistent with ATP supply influencing rubisco activation). Although evidence verifies that the activation state of rubisco declines under low light and RuBP levels remain high, the

Farquhar model, with its faulty assumption of RuBP limitation at low light, nevertheless behaves realistically. According to P.C. Harley (1990, personal communication), the Farquhar model, Eq. [3] of Evans and Farquhar (Ch. 1 of this book), and their Eq. [9] also predict that the slopes of the A/C_i curves are light dependent below 350 μmol PPFD m^{-2} s^{-1}.

In the original Farquhar model, the transition to high CO_2 concentration is proposed to be characterized by RuBP-concentration limitation of assimilation. More recent research suggests that the transition to the "RuBP-limiting" region of the C_i response curve is not really RuBP-limited, but the result of less activation of rubisco via effects on rubisco activase and carbamylation. Since rubisco activation by rubisco activase is dependent on ATP supply (Streusand & Portis, 1987), and activation is decreased in the presence of ATP-consuming reactions (Machler & Nosberger, 1984), it is reasonable to assume that, as C_i increases to high levels, more of the energy would go toward CO_2 fixation and reduction, thus reducing the activation state of rubisco at high CO_2 concentration, as observed by Sage et al. (1988).

Taylor and Terry (1984) proposed that Eq. [7], describing the "CO_2-limited" region, should be modified to mimic the effects of light activation of rubisco as follows:

$$A = [C_iM/\ (C_iM + K_{cm})]\ V_{max}(C_i - \Gamma^*)/ [C_i + K_c(1 + O/K_o)] - R_d \quad [8]$$

The term $[C_iM/(C_iM + K_{cm})]$ was suggested by Taylor and Terry (1984) to represent the degree of activation of rubisco, where M represents Mg^{2+} concentration and K_{cm} represents the affinity constants for CO_2 and Mg^{2+}. A more consistent approach that would also account for energy-limiting effects at high C_i would be to use an "activation term", asymptotically dependent on ATP level, that accounts for concurrent use of energy for CO_2 assimilation, photorespiration, and other cellular processes. Taylor and Terry (1986) demonstrated that rubisco activation appears to be quite directly related to the rate of photosynthetic electron transport, whether caused by varying irradiance or by varying the electron-transport capacity. When electron-transport capacity (light-harvesting system) was decreased by Fe deficiency, the fraction of rubisco activated was proportionately less, although total rubisco protein was not affected.

In spite of problems with assumptions regarding RuBP levels and rubisco activation, the A/C_i approach of Farquhar et al. (1980) has proven very useful and was used by four of our authors (Evans & Farquhar, Ch. 1; Harley & Tenhunen, Ch. 2; Gutschick, Ch. 4; Norman & Arkebauer, Ch. 5).

Light Effects on Mitochondrial Respiration

The above equations predict leaf net assimilation rate, i.e., predict gross assimilation rate and then subtract leaf R_d. Normal mitochondrial respiration is believed to continue in the light, but at a lower rate dependent on PPFD (Kirschbaum & Farquhar, 1987; Müller, 1986; Peisker et al., 1983). The R_d is gradually reduced as PPFD increases from 0 to 100 μmol m^{-2} s^{-1}. A consequence of the gradual reduction in R_d is that the apparent CO_2 compensation point is increased sharply below 100 μmol m^{-2} s^{-1} (Peisker et al., 1983) and the apparent Q_E is greater in this region (Kirschbaum & Farquhar, 1987).

Acock Approach for Predicting Gross Photosynthesis from External Carbon Dioxide Concentration

Acock (Ch. 3 in this book) uses the following equations to predict gross photosynthesis (P_g) and photorespiration (R_p) vs. ambient CO_2 and O_2 concentrations:

$$P_g = \alpha_m I \tau C_a / (\alpha_m I + \tau C_a) \quad [9]$$

$$R_p = \alpha_m I \Omega O / (\alpha_m I + \tau C_a) \quad [10]$$

where α_m is leaf light-utilization efficiency
τ is leaf conductance to CO_2 transfer
Ω is leaf conductance to O_2 transfer
I is incident PPFD
C_a is ambient CO_2 concentration
O is ambient O_2 concentration

Net leaf assimilation, then, is predicted from

$$A = P_g - R_p - R_d. \quad [11]$$

These equations thus address the combined effects of PPFD and CO_2 and O_2 concentrations. The combined equations have been described previously by Charles-Edwards (1981). Leaf response to C_a was assumed to have two features: (i) a CO_2-limiting region in which assimilation increases linearly in response to CO_2 concentration, and (ii) a CO_2-saturating region (P_{max_c}), in which further increases in CO_2 concentration fail to increase photosynthesis. Leaf response to PPFD likewise has a light-limiting region (α_m) and a light-saturating region (P_{max_l}). Both P_{max_l} and P_{max_c} are affected by temperature and light history, which influence leaf thickness, N content, and other factors.

Response to Temperature

The processing time per photon, electron, or CO_2 molecule is certainly temperature dependent, but this was discussed only by Harley and Tenhunen (Ch. 2 of this book). Temperature effects on J_{max} or V_{max} have been described by Farquhar et al. (1980), Harley et al. (1985), and Harley and Tenhunen (Ch. 2 of this book) in terms of activation energy, energy of deactivation, an entropy term, and a scaling constant. The latter authors describe dependencies of the specificity ratio, K_c, K_o, and R_d by exponential functions, which use only the activation energy and a scaling constant. Acock (Ch. 3 of this book) assumed a linear response of CO_2-saturated assimilation rate to temperature in the range of 5 to 35° C for soybean [*Glycine max* (L.) Merr.]. None of the authors described the solubility of CO_2 or O_2 in aqueous phase as a function of temperature, but dealt with gas-phase concentrations of the gases. As a result, the apparent K_c and K_o constants for the Michaelis-Menten competition kinetics are highly sensitive to temperature and increase exponentially with temperature (see Harley et al., 1985; Harley & Tenhunen, Ch. 2 in this book).

Temperature effects may be important under field conditions, although Sinclair (Ch. 6 in this book) makes a good point that crop biomass accumulation is not as sensitive to regional differences in temperature as one would presume from looking at leaf assimilation response to temperature. Temperature has a much greater effect on J_{max} or P_{max} (or V_{max} of enzymes) than it does on Q_E (Harley et al., 1985). Harley et al. (1985) showed that Q_E of soybean (on an incident-light basis in saturating CO_2 concentration and low O_2) averaged 0.0623 mol CO_2 mol^{-1} photosynthetically active photons for temperatures from 20 to 40° C, with a very slight decline from 25 to 40° C, but a larger reduction below 15° C. Assuming 80% absorption, this compares well with the Q_E of 0.073 on an absorbed-PPFD basis reported by Ehleringer and Björkman (1977) under CO_2-saturating conditions. By contrast, when measured under ambient CO_2 and O_2 levels, there is a steady decline in Q_E as temperature increases from 14 to 38° C (Ehleringer & Björkman, 1977; Ehleringer & Pearcy, 1983). Likewise, it is important to realize that the P_{max} reported by Harley et al. (1985) was measured under CO_2 saturation, light saturation, and low O_2 concentration. We suspect that apparent light-saturated P_{max} measured under normal field conditions (ambient CO_2 and O_2 concentrations), would have a much broader temperature optimum from 20 to 35° C, as demonstrated by the simulations of Farquhar et al. (1980).

Coupling of Assimilation Rate and Stomatal Conductance

The linkages between assimilation rate, stomatal conductance, and transpiration rate are critical factors determining productivity of vegetation in water-limited environments. Three of the modeling approaches (Harley & Tenhunen, Ch. 2; Gutschick, Ch. 4; and Norman & Arkebauer, Ch. 5) include a coupling of stomatal conductance to assimilation rate via the C_i/C_a ratio. One approach, used by Gutschick and by Norman and Arkebauer, is to assume the ratio C_i/C_a is a constant. An alternate view is that some species may hold constant C_i, irrespective of C_a (de Wit, 1978, p. 43).

Harley and Tenhunen, following the approach of Ball et al. (1987), describe stomatal conductance (g_s) as a function of relative humidity (h_s),

CO_2 concentration at the leaf surface (C_s), leaf assimilation rate, and a constant (k) representing stomatal sensitivity to these factors:

$$g_s = k\,A\,h_s/C_s \qquad [12]$$

As described above, stomatal conductance is thus dependent on A. Leaf assimilation rate is, in turn, dependent on g_s via the latter's effect on C_i, according to the equation

$$C_i = C_a - 1.56\,A\,\mathrm{ATM}/g_s \qquad [13]$$

where ATM is atmospheric pressure and the value 1.56 is the ratio of diffusivities of water vapor and CO_2 in air. Equation [13] must hold in either case, whether holding the ratio C_i/C_a constant, or fixing a constant value for k. Since A is dependent on C_i, and C_i is partially dependent on A, the values for A and C_i must be solved iteratively to give a value of C_i compatible with both the conductance predicted by the equation above and with assimilation rate predicted by the earlier C_i-dependent assimilation models (see comments about this iterative process by Harley & Tenhunen, Ch. 2, and by Gutschick, Ch. 4). The iterative process becomes more complex with the addition of an energy-balance loop described by Norman and Arkebauer, Ch. 5, and by Gutschick. It would become even more complex if root water uptake, soil conductivity, and root conductivity to water flow were to create limits to transpirational water loss.

Solving Parameters from Experimental Data

Harley and Tenhunen describe in considerable detail a systematic approach by which the various parameters can be solved with nonlinear regression from data on leaf assimilation. To derive the full set of coefficients, they needed assimilation response to light under saturating CO_2 concentration, and assimilation response to C_i under saturating light at a range of temperatures. They assume that α (Q_E under CO_2-saturated conditions), K_c, K_o, and the specificity factor are constants, primarily based on values found repeatedly in the published literature. One may appropriately ask whether solving for all of these parameters is needed to predict assimilation. More importantly, independent validation is not likely to be possible because the next leaf sampled has a different P_{ml} (or V_{max} of rubisco). Thus, to be independent from the need to calibrate, we need more independent ways to predict the saturated electron-transport rate and V_{max} (of rubisco) from simpler leaf traits (e.g., leaf N concentration, specific leaf mass, chlorophyll concentration, recent light history of the leaf) that could be state variables in whole-crop models.

SCALING UP FROM LEAF TO CANOPY ASSIMILATION

The basic problem in scaling photosynthesis models from single leaves to canopies is prediction of light distribution among the leaf elements within the canopy. The distribution of intercepted light among leaf elements is usually rather complex, even for the simple canopies produced by crop monocultures. The light sources consist of direct rays from the sun, diffuse skylight, and light that penetrates to lower levels after reflection or transmission by upper leaves. Each of these sources change as the sun's position changes (as a function of latitude, day of the year, and hour) and as atmospheric conditions change. In the lower latitudes and with low humidity and clear skies, skylight amounts to about 20% of the total solar flux, but that fraction increases dramatically with increases in latitude and humidity.

Canopy structure adds additional complexity. Leaf elevation angle and solar elevation cause the irradiance on leaves exposed to direct light to vary with the cosine of the incidence angle. In addition, the sun has a single azimuth at any time, whereas leaves vary in their azimuth (e.g., uniformly to all compass points or oriented in specific directions). Within a single horizontal layer of leaves, the distribution of individual leaves may be displayed in patterns ranging from quite regular to strongly clumped. Additional subtleties impacting photosynthesis arise through variations in microclimate (temperature, humidity, and CO_2 concentration as well as light) and leaf properties with depth in the canopy.

The questions we address here relate to practical ways of simplifying canopy and light-distribution geometry for use in crop models.

Simple vs. Complex Approaches

There are several approaches for predicting instantaneous canopy assimilation, using the leaf assimilation traits described above. The first approach is merely to multiply leaf assimilation (in full sun) by the leaf area index (LAI). This approach is valid only for small LAIs (< 1.0) because it assumes all leaves are fully exposed to the sun. It suffers serious overestimation problems as LAI increases and as light flux increases, because it does not account for mutual shading within the canopy. A second approach, usable only for full canopies, could be termed the *big-leaf approach*, which assumes the canopy is one continuous sheet of leaf area with no gaps. This approach underestimates canopy assimilation. It fails to address variations in leaf display and ignores transmittance and reflection of light to lower shaded leaves where small fluxes are utilized with high efficiency. A third approach is an analytical solution to canopy assimilation as proposed by Acock et al. (1978). Used with caution, the analytical approach has merit. A fourth, simplified approach based on sunlit and shaded LAI considers projection angle of the leaves normal to the solar path, leaf transmittance to light, and canopy reflectance. This approach is used by Sinclair (Ch. 6

of this book), Sinclair and Horie (1989), Boote and Jones (1987), Boote et al. (1988), and Norman (1982). It is more than a big-leaf model because it considers assimilation by a second category of shaded leaves. A fifth, optimum approach involves numerical integration, in which direct beam, skylight, and transmitted light are allowed to be absorbed by successive leaf layers, and each layer has the potential to have different leaf display and different assimilation traits (Denison & Loomis, 1989; Duncan et al., 1967; Norman, 1979, 1986; Norman & Campbell, 1983; Ross, 1981). This approach becomes complex when all relevant energy and mass-transfer processes are considered, but the predictions can be excellent.

Light Absorbance, Reflectance, and Transmittance

In its simplest form, a crop canopy can be thought of as one hypothetical big leaf, i.e., one unit of LAI with no random gaps. For single leaves and for such a big-leaf example, one must consider absorptance, reflectance, and transmittance of light. Up to 10% of incident PPFD is reflected and up to 10% is transmitted through individual leaves; therefore 80 to 90% is absorbed. Scattering (transmittance and reflectance) of direct-beam irradiance to lower leaves and penetration of diffuse skylight are important to canopy photosynthesis in two ways. In a full crop canopy (e.g., LAI = 5, distributed uniformly in a horizontal plane), the lower, shaded leaves absorb skylight that escapes the upper leaves as well as 10 to 15% of the incident PPFD that is scattered by the sunlit leaves. Moreover, the light flux absorbed by lower leaves is generally small, with the result that lower leaves operate closer to their maximum Q_E. Because of these two features, assimilation by real canopies is greater than that predicted for a big-leaf model consisting of one contiguous unit layer of LAI. The second curve in Fig. 7–2 represents the

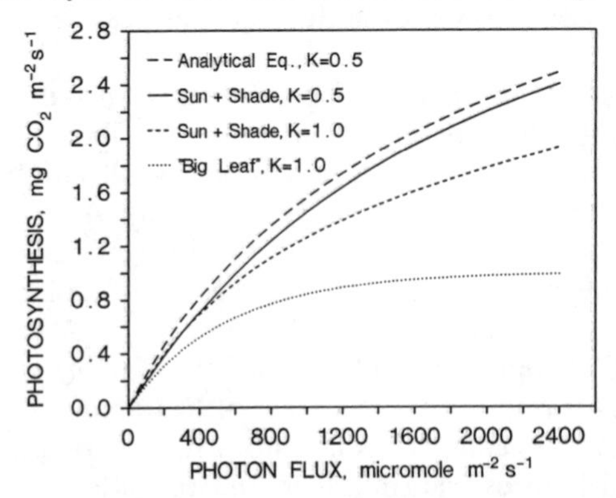

Fig. 7–2. Assimilation response to increasing PPFD by: (a) single horizontal leaves (same as canopy rate of horizontally angled sunlit LAI or big-leaf canopy); (b) canopy rate for horizontally angled sunlit-plus-shaded LAI (K_d = 1, horizontal leaves, random); (c) canopy rate for spherically angled sunlit-plus-shaded LAI (K_d = 0.5, spherical leaf-angle distribution, random); (d) canopy rate using the analytical equation of Acock et al. (1978) with K_d = 0.5. Simulations assume LAI = 5.0, σ = 0.20, Q_E = 0.0524 mol mol^{-1}, P_{max} = 1.0 mg CO_2 m^{-2} s^{-1}, and sun directly overhead.

total assimilation of sunlit plus shaded leaves, for a canopy of LAI = 5.0 with horizontal leaves in which lower leaves receive 20% of the PPFD resulting from either scattering of direct beam by sunlit leaves or penetration of skylight to lower leaves. This provides a considerable advantage over assimilation by sunlit leaves only (lower curve, Fig. 7–2), which would be comparable to photosynthesis of a big-leaf model.

Random Distribution of Leaf Area and Light Interception

Because leaves in crop canopies are not contiguous, there are gaps between adjacent leaves through which light passes further down in the canopy. Assuming the gaps are randomly distributed horizontally, the area of direct-beam irradiance penetrating to any depth in the canopy is an exponential function of the cumulative LAI from the top of the canopy:

$$I = I_o\, exp(-K_{\mathrm{d}}\ \mathrm{LAI}) \quad [14]$$

where I and I_0 are the areas of direct-beam fluxes to horizontal receivers at points within and above the canopy, respectively (Monsi & Saeki, 1953; Loomis & Williams, 1969). In the area context, the direct-beam extinction coefficient (K_{d}) is the ratio of horizontally projected shadow area per unit land area per unit leaf area. The shadow projection of leaves depends on leaf angle and solar elevation angle (β). The value of K_{d} is 1.0 only when all leaves are distributed randomly in the horizontal plane and are perpendicular to the direct beam with solar elevation of 90°. Under these conditions, a canopy of LAI = 1 intercepts 63% of the incident direct irradiance. The value of K_{d} varies by the factor $1/\sin(\beta)$ as the solar position changes. Variations in leaf angle also have a large effect on K_{d} because less light is intercepted when leaves are displayed obliquely to the direct beam. An assumption of randomness approximates most crop canopies. Departures are noticeable in clumped (e.g., young stands of row crops), regular (e.g., clover [*Trifolium* spp.]), and solar-tracking (legumes, cotton [*Gossypium* spp.]) communities. In addition to K_{d}, total light extinction by a crop canopy requires consideration for scattering (reflectance and transmittance) as was done by Duncan et al. (1967) and Goudriaan (1977, 1982). For these reasons, experimentally determined values for total light extinction would not necessarily be the same as K_{d}. Readers are referred to Loomis and Williams (1969), Ross (1981), and Campbell and Norman (1989) for further information on light interception relative to leaf angle, leaf pattern, LAI, and solar elevation.

Effects of Leaf Angle and Solar-Beam Angle on Photosynthesis

Diurnal variation in solar elevation and variation in leaf angle create situations where the sunlit leaves are obliquely illuminated and thus do not receive full direct-beam irradiance. This differs in important ways from the assumption that leaves are either horizontal or perpendicular to the direct-beam irradiance. In the latter two situations, a smaller fraction of upper leaves would operate at irradiances close to their light-saturation point, while a larger fraction of the leaf area would receive only low irradiance. Typically, most leaves in crop canopies are at some angle less than perpendicular to the solar beam and the average irradiance per unit leaf area is less, thus allowing more leaves to operate where the incremental quantum efficiency is greater. The result of leaf angle–solar beam angle consideration is that canopy assimilation is greater, especially at midday, than would be predicted by a model assuming horizontal leaves. This benefit of increasing leaf angles on canopy assimilation has been simulated by Duncan (1971) and by Loomis and Williams (1969). The curves in Fig. 7–2 illustrate the canopy-assimilation advantage at midday gained by a more upright leaf-angle distribution ($K_d = 0.5$) vs. a horizontal leaf-angle distribution ($K_d = 1.0$)

Simple Models to Predict Canopy Assimilation Response to Light

In order to evaluate the effects of random horizontal leaf arrangement and leaf angle relative to solar angle (via K_d), we will introduce two simple models: an analytical solution to canopy assimilation (Acock et al., 1978), and a simplified approach based on sunlit and shaded LAI (Boote & Jones, 1987; Sinclair & Horie, 1989; Sinclair, Ch. 6 of this book).

First, let us consider photosynthesis by sunlit leaves. The amount of sunlit LAI (LAI_{sun}) is described analytically from K_d and total LAI:

$$LAI_{sun} = (1/K_d)\,[1 - \exp(-K_d\, LAI)] \qquad [15]$$

If $K_d = 1$ (perpendicular to solar beam), then sunlit LAI approaches 1 as LAI goes to infinity and assimilation by LAI_{sun} would be the same as the big-leaf model. If $K_d < 1$, say 0.5, then the amount of sunlit LAI would approach 2 as LAI goes to infinity. The LAI_{sun} at midday for an LAI of 5.0 is 0.99 and 1.84 for K_d of 1.0 and 0.5, respectively. The incident PPFD on LAI_{sun} would be $K_d \times$ PPFD. After accounting for light scattering (σ) caused by reflectance and transmittance, the absorbed PPFD would be $(1 - \sigma) \times K_d \times$ PPFD, and assimilation rate of sunlit leaves (P_{sun}) would be

$$P_{sun} = P_{max}\,\{1 - \exp[-Q_E\, K_d\, (1 - \sigma)\, PPFD/P_{max}]\} \qquad [16]$$

Next, let us consider photosynthesis by shaded leaves. Total canopy photosynthesis (P_{can}) will be increased to the extent that skylight and direct beam scattered and transmitted through the upper leaves is absorbed by lower, shaded leaves. Shaded LAI (LAI_{shd}) is defined as any LAI that is not sunlit. Mathematically, LAI_{shd} equals total LAI minus LAI_{sun}. In a typical situation, about 10 to 20% of the direct beam is scattered through sunlit leaves to shaded leaves. We will assume $\sigma = 0.20$ also includes some diffuse skylight. As an approximation, Boote and Jones (1987) assumed that LAI_{shd} intercepted this scattered light using the same K_d, and that all shaded LAI shared equally in the use of this scattered PPFD, once intercepted. Thus, PPFD incident and available for photosynthesis per unit shaded LAI is

$$\mathrm{PPFD_{shd}} = \sigma\ \mathrm{PPFD}\ [1 - \exp(-K_d\ \mathrm{LAI_{shd}})]/\mathrm{LAI_{shd}} \qquad [17]$$

Thus, assimilation by shaded leaves (P_{shd}) is

$$P_{shd} = P_{max}\ [1 - \exp(-Q_E\ \mathrm{PPFD_{shd}}/P_{max})]. \qquad [18]$$

Total-canopy assimilation summed across both categories of leaves is

$$P_{can} = P_{sun}\ \mathrm{LAI_{sun}} + P_{shd}\ \mathrm{LAI_{shd}} \qquad [19]$$

Simulations in Fig. 7-2 illustrate the respective benefits to P_{can} attributed to light scattering to shaded leaves and attributed to leaf angle of sunlit leaves when solar elevation is 90°. The difference between the two lower curves illustrates the benefits of light scattering-transmittance to shaded leaves for a canopy having an LAI of 5.0 and a K_d of 1.0 (horizontal leaves) compared with a big-leaf canopy (horizontal leaves) that does not consider light scattering. The lower curve also describes single-leaf photosynthesis response to PPFD. The third curve highlights the benefits of more upright leaf angle ($K_d = 0.5$) in addition to light scattering to shaded leaves. Upright leaf angle actually reduces total light interception, but it exposes more leaves to direct beams of lower average irradiance where the quantum efficiency is greater. Predictions with the sunlit and shaded LAI approach are compared in Fig. 7-2 with predictions by the analytical equation for canopy assimilation by Acock et al. (1978). Their analytical equation is the integral of leaf photosynthesis (Michaelis-Menten equation) over LAI depth as PPFD is absorbed:

$$P_{can} = \frac{P_{max}}{K_d} \ln \left| \frac{(1 - \sigma)\ P_{max} + Q_E\ K_d\ \mathrm{PPFD}}{(1 - \sigma)\ P_{max} + Q_E\ K_d\ \mathrm{PPFD}\ \exp(-K_d\ \mathrm{LAI})} \right| \qquad [20]$$

This analytical equation gives comparable, but slightly higher, predictions compared with the sunlit-plus-shaded LAI approach when the same values for K_d and σ are used (Fig. 7-2). Given inputs of field measurements of leaf P_{max}, K_d, and LAI, the sunlit-plus-shaded LAI photosynthesis model has adequately predicted total canopy assimilation respnse to PPFD measured on closed canopies of soybean, peanut (*Arachis hypogaea* L.), and common bean (*Phaseolus vulgaris* L.) (Boote & Jones, 1987; Boote et al., 1988). Predictions with the sunlit-plus-shaded LAI approach have generally been closer to observed total-canopy photosynthesis than predictions with the analytical equation of Acock et al. (1978).

The simplified sunlit vs. shaded LAI approach was adopted with minor modifications by Sinclair for his chapter (Ch. 6) and was used by Boote et al. (1988) with further modifications to create a hedgerow assimilation model.

Including the effects of CO_2 concentration or temperature in the equations appears feasible because the equations essentially use single-leaf photosynthesis multiplied by sunlit or shaded LAI. As presented above, these equations do not consider effects of vertical profiles in leaf traits or effects of hedgerow canopies.

Vertical Profiles in Leaf Traits, Light, and Carbon Dioxide Concentration

Another aspect of crop canopies of concern to modelers of canopy assimilation is how physical inputs of light, CO_2 concentration, temperature, and wind speed vary with depth in the crop canopy. If the absorption of total solar radiation by respective layers of a leaf canopy has been simulated, then leaf energy balance can be calculated if the wind speed profile and leaf conductance to water-vapor loss are also simulated. Both Norman (Ch. 5) and Gutschick (Ch. 4) use canopy-assimilation models that consider leaf energy balance in the respective layers and consider vertical profiles in light absorbed, air temperature, and wind speed. Vertical profiles in CO_2 concentration also may have a minor effect, since a 5 to 10% decrease in CO_2 concentration in the middle of a short canopy has been observed at midday vs. the concentration above the canopy (Lemon, 1969) and up to 25% decrease has been observed in corn (*Zea mays* L.) on a still day (Chapman et al., 1954).

Leaf characteristics, including P_{max}, dark respiration (R_d), protein concentration, and specific leaf mass (SLM), also vary with depth in the canopy. Changes in these characteristics are caused primarily by variations in nutrition and by gradients in irradiance and leaf age. Upper leaves are exposed to greater cumulative irradiance, whereas lower leaves may lose protein and photosynthetic capacity as a result of shading and aging. The values of P_{max} and R_d are generally associated with each other, and are approximately proportional to the N content per unit leaf area (also proportional to SLM) and to the cumulative light history (which may also affect N content and SLM). (See Acock [Ch. 3 in this book] for an approach to predicting leaf carboxylation conductance [proportional to P_{max}] from the "past week's" irradiance receipts.)

Importance of Vertical Gradients in Leaf Traits to Canopy Assimilation

The effect of vertical gradients in P_{max} and R_d can be important to the prediction of canopy assimilation, depending on the type of canopy-assimilation model. For general work and nonlayered canopy models, the P_{max} from upper sunlit leaves can usually be accepted as sufficient throughout the canopy, because shaded leaves operate mainly in low light and thus operate close to their region of higher quantum efficiency where Q_E is important but P_{max} is less important. This argument holds to the extent that lower leaves receive only transmitted and diffuse light. The value selected for P_{max} for lower leaves becomes more important for layered models that consider the penetration of sunflecks to shaded leaves, as is the case for the

Norman and Arkebauer model (Ch. 5). In such layered models, one must decide whether lower leaves receiving direct-beam irradiance can use it efficiently, or whether P_{max} should be adjusted vertically. Meister et al. (1987) found that accurate prediction of photosynthesis by oak (*Quercus coccifera*) trees required adjustment of P_{max} with depth, but not of Q_E.

For layered canopy models that attempt to predict net canopy assimilation (P_{net}) based on R_d input functions, problems may also develop because R_d varies strongly with canopy depth. Duncan et al. (1967), using a numerical "layered" simulator for P_{net} prediction, clearly showed that using a fixed R_d for all leaves caused incorrect prediction of the curvature of canopy assimilation vs. LAI. Profile adjustments of R_d are, thus, very important for layered models that predict P_{net} upon subtraction of R_d. Norman and Arkebauer (Ch. 5 of this book) predict net assimilation with a numerical approach; they consider vertical gradients in environmental inputs and vertical adjustments in P_{max} and R_d.

The question of vertical gradients in R_d is not applicable to canopy models that predict gross assimilation (assimilation goes to zero at zero PPFD). Such canopy-assimilation models are typically used in crop-growth models that predict R_d of tissues in other subroutines.

The numerical simulators for predicting assimilation by layered canopies easily incorporate the vertical gradients in environmental inputs and leaf characteristics. They clearly represent the most versatile and powerful means for predicting canopy photosynthesis and are yet to be matched by simpler or analytical models. The big-leaf and multiply-by-LAI models cannot handle vertical gradients at all. The sunlit-vs.-shaded-LAI approach can be adapted to consider different environmental inputs and different leaf characteristics (P_{max} and R_d) for the two respective layers. For example, lower leaves may be assumed to receive only transmitted light plus a fraction of diffuse skylight. They may be assumed to have different wind speed, temperature, and ambient CO_2 concentration inputs as well. The P_{max} and R_d for the shaded LAI could be decreased if desired in the sunlit–shaded LAI approach. For example, decreasing the P_{max} of shaded leaves by one-half decreased simulated P_{can} by 4 to 6% at times of high irradiance, but by only 2 to 3% when integrated throughout a whole day. The analytical solutions for canopy assimilation (Acock et al., 1978; Acock, Ch. 3 of this book) ordinarily do not address vertical gradients in inputs other than light. Nevertheless, Acock (Ch. 3 of this book) attempted to consider vertical gradients in leaf conductance by creating a numerical layering within his canopy-assimilation model.

Row-Crop Canopies and Nonrandom Horizontal Distribution of Leaf Area Index

For the above simple examples, we assumed that leaves are randomly positioned horizontally and vertically, thus allowing the typical exponential extinction of light in the crop canopy. For actual crop canopies, the crop is frequently planted in rows and foliage is confined within a hedgerow. Even if plants are placed in equidistant spacing, leaves may be clumped about the main axis of the plant. The leaf area then is nonrandomly distributed horizontally, allowing greater light capture in clumped regions and greater light penetration to the soil where gaps exist within the canopies and between adjacent rows. If one looks at the canopy hedgerows at midday with the sun directly overhead, the impression may be that considerable light is not captured. Nevertheless, effect of time of day and time of year on solar angle causes considerably greater capture at times when the solar elevation is $< 90°$. It is possible to simulate the light interception by a hedgerow using three-dimensional geometry if one considers hedgerow dimensions, time of day, and time of year (Allen, 1974; Gijzen & Goudriaan, 1989). (See discussion by Acock [Ch. 3 of this book] and the hedgerow model presented below.)

Assimilation by Hedgerow Canopies

A model to predict assimilation of row-crop canopies was developed by Boote et al. (1988, 1989) based on a simplification of the hedgerow approach of Gijzen and Goudriaan (1989). The approach considers sunlit and shaded LAI similar to that described by Boote and Jones (1987) and used by Sinclair (Ch. 6 of this book). The shadow projected by the canopy is computed as a function of canopy height (*H*), canopy width (*W*), time of day, day of year, latitude, and row azimuth. The LAI is allowed to intercept light only in the shadow-projection zone. Acock (Ch. 6) described a similar approach for light interception by crop rows. The canopy envelope (Fig. 7-3) is defined in the plane perpendicular to the row direction and is assumed to have a height, a width, and an effective curvature at the top and bottom of the canopy that is equivalent to a half circle with radius equal to one-half width. Let ϕ be the effective angle from the edge of the shadow

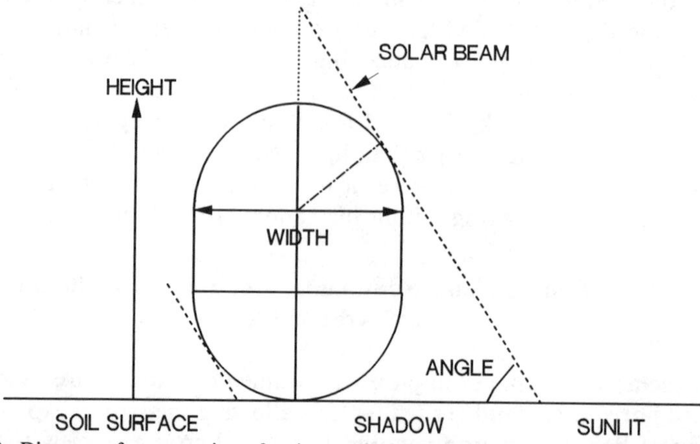

Fig. 7-3. Diagram of cross section of an incomplete row-crop canopy, assuming canopy height (*H*), canopy width (*W*), and a curvature at the top and bottom of the canopy equivalent to a half circle with radius of $W/2$.

to the top of the canopy (in the plane perpendicular to the row), it can be predicted from row direction, solar azimuth, and β. Solar azimuth and β are computed from standard equations as a function of latitude, day of year, and time of day (Goudriaan, 1982).

The length of shadow (Z_{sh}) cast by the canopy in the plane perpendicular to the row is computed as

$$Z_{sh} = [H - W + W/\cos(\phi)]/\tan(\phi). \quad [21]$$

Light interception, photosynthesis, and LAI are restricted to the fraction of the soil surface shaded by the canopy (f_{sh}), which is a function of (Z_{sh} and row spacing (R), although the leaf area density in the shaded zone is increased according to LAI/f_{sh}. The f_{sh} is computed as Z_{sh}/R, restricted not to exceed 1.0:

$$f_{sh} = \min(1.0, Z_{sh}/R) \quad [22]$$

For simplicity, a spherical leaf-angle distribution was assumed, where the average projection of leaves relative to the direct beam path is 0.5. The extinction coefficient (K_d) also varies with β, because β affects the effective pathlength through the canopy.

$$K_d = 0.5/\sin(\beta) \quad [23]$$

Experience has shown, however, that a spherical leaf-angle distribution results in an underestimate of light interception at midday for soybean and peanut and may not be a good assumption for these crops because both crops tend to be regular, planophile, and partially solar-tracking types.

The sunlight and shaded LAI are computed as

$$\mathrm{LAI}_{sun} = (f_{sh}/K_d)\,[1 - \exp(-K_d\,\mathrm{LAI}/f_{sh})] \quad [24]$$

$$\mathrm{LAI}_{shd} = \mathrm{LAI} - \mathrm{LAI}_{sun} \quad [25]$$

Total incoming PPFD is distributed into a direct component (PPFD_{dir}) and diffuse component (PPFD_{dif}) dependent on β as follows:

$$\mathrm{PPFD}_{dif} = \{1.00 - \exp[-0.20/\sin(\beta)]\}\ \mathrm{PPFD} \quad [26]$$

$$\mathrm{PPFD}_{dir} = \mathrm{PPFD} - \mathrm{PPFD}_{dif} \quad [27]$$

The PPFD absorbed by sunlit and shaded LAI is computed as described by Spitters (1986): a proportion of the direct-beam PPFD is converted to diffuse within the canopy by scattering processes. The scattering coefficient (σ) includes transmittance and reflectance down through the foliage. Reflectance (δ) from the canopy to the sky depends on σ and β.

$$\delta = [1 - (1 - \sigma)^{1/2}]/[1 + (1 - \sigma)^{1/2}]\ 2/[1 + 1.6 \sin(\beta)] \quad [28]$$

$$A_{\mathrm{dir,dir}} = (1 - \sigma)\ \mathrm{PPFD_{dir}}\ [1 - \exp(-K_{\mathrm{d}}\ \mathrm{LAI}/f_{\mathrm{sh}})] \quad [29]$$

$$A_{\mathrm{dir,tot}} = (1 - \delta)\ \mathrm{PPFD_{dir}}\ (1 - \exp(-K_{\mathrm{d}}\ (1 - \sigma)^{1/2}\ \mathrm{LAI}/f_{\mathrm{sh}})] \quad [30]$$

$$A_{\mathrm{dir,dif}} = A_{\mathrm{dir,tot}} - A_{\mathrm{dir,dir}} \quad [31]$$

Absorption of skylight (A_{dif}) is computed following an approach from Goudriaan (1977, p. 59–63) that uses the path width (alley between hedges), the height and width of the hedgerow, LAI, and a diffuse extinction coefficient of 0.8. Goudriaan's approach assumed that diffuse irradiance originates from a uniformly overcast sky.

The PPFD absorbed by the sunlit and shaded classes of leaves is then computed. The average flux of PPFD absorbed by the shaded leaves ($\mathrm{PPFD_{shd}}$) comes from absorbed skylight (A_{dif}) and from direct-beam irradiance converted to diffuse within the canopy ($A_{\mathrm{dir,dif}}$). The flux absorbed by sunlit leaves ($\mathrm{PPFD_{sun}}$) includes $\mathrm{PPFD_{shd}}$ plus direct-beam irradiance.

$$\mathrm{PPFD_{shd}} = A_{\mathrm{dif}}/\mathrm{LAI} + A_{\mathrm{dir,dif}}/\mathrm{LAI} \quad [32]$$

$$\mathrm{PPFD_{sun}} = \mathrm{PPFD_{shd}} + (1 - \sigma)\ K_{\mathrm{d}}\ \mathrm{PPFD_{dir}} \quad [33]$$

$$P_{\mathrm{sun}} = P_{\mathrm{max}}\ [1 - \exp(-Q_{\mathrm{E}}\ \mathrm{PPFD_{sun}}/P_{\mathrm{max}})] \quad [34]$$

$$P_{\mathrm{shd}} = P_{\mathrm{max}}\ [1 - \exp(-Q_{\mathrm{E}}\ \mathrm{PPFD_{shd}}/P_{\mathrm{max}})] \quad [35]$$

$$P_{\mathrm{can}} = P_{\mathrm{sun}}\ \mathrm{LAI_{sun}} + \mathrm{P_{shd}}\ \mathrm{LAI_{shd}} \quad [36]$$

Simulated Photosynthesis of Closed vs. Hedgerow Canopies

Instantaneous canopy assimilation with the hedgerow model is illustrated in Fig. 7–4 to evaluate effects on photosynthesis of different row spacings with incomplete canopy cover. Simulations were conducted using LAI = 3 with row spacings of 0.5, 1.0, and 2.0 m. The canopy was assumed to be 0.7 m tall and 0.7 m wide with a spherical leaf-angle distribution [$K_{\mathrm{d}} = 0.5/\sin(\beta)$], north–south rows, $\sigma = 0.20$, $Q_{\mathrm{E}} = 0.0524$ mol mol^{-1}, and $P_{\mathrm{max}} = 1.0$ mg CO_2 m^{-2} s^{-1} for simulations on 1 July at 30 °N latitude. For the closed-canopy situation (0.5-m rows), the hedgerow model predictions were within 1% of the simple sunlit–shaded LAI model presented above (minor differences relate to different approaches for distribution and capture of diffuse irradiance and reflectance). When row spacing was increased to 1.0 m, canopy assimilation was reduced only during midday, when direct-beam irradiance increasingly penetrated to the soil. With a 2.0-m row spacing, assimilation was lower for much of the day. It is important to emphasize that LAI was held constant per unit land area with higher density

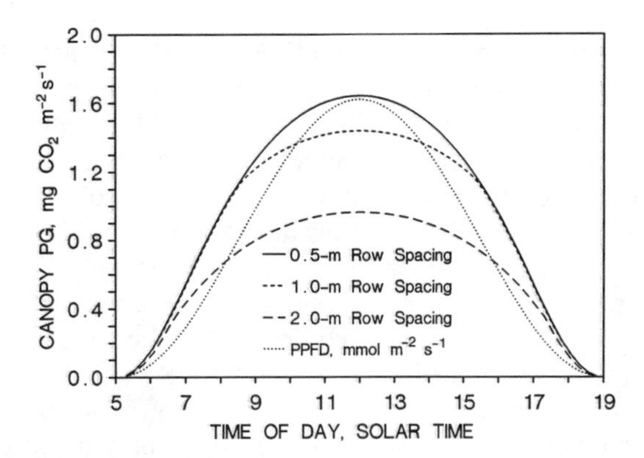

Fig. 7–4. Instantaneous canopy assimilation rate predicted by the hedgerow model throughout the day to illustrate the effect of row spacing (incomplete canopy cover) on canopy photosynthesis. Simulations were conducted using LAI = 3 with row spacings of 0.5, 1.0, and 2.0 m. The canopy was assumed to be 0.7 m tall and 0.7 m wide, in north–south rows, with a spherical leaf-angle distribution [0.5/sin(β)]. Simulations assumed daily PPFD = 45 mol m^{-2} d^{-1}, $\sigma = 0.20$, $Q_E = 0.0524$ mol mol^{-1}, and $P_{max} = 1.0$ mg CO_2 m^{-2} s^{-1} for simulations on 1 July at 30° N latitude.

in the canopy envelope as row spacing was increased. Although canopy cover at midday was only 70 and 35% for the 1.0- and 2.0-m row spacings, respectively, canopy assimilation was not comparably reduced, partly because of diffuse-light interception and greater leaf area density in the canopy envelope.

This simplified approach for hedgerow photosynthesis accounts for most of the effect of row spacing on canopy photosynthesis. Additional refinements such as leaf-angle variation, effects of plant spacing in the row, and PPFD reflectance from the soil are included in the version developed by Boote et al. (1988, 1989). The hedgerow assimilation model has performed well in predicting field-measured gross canopy assimilation for soybean and peanut in various row spacings where inputs were measured leaf P_{max}, measured LAI, measured row spacing, canopy height, canopy width, day of year, time of day, and PPFD (Boote et al., 1988).

Going from Instantaneous to Daily Canopy Assimilation

To compute canopy assimilation throughout the day under field conditions, instantaneous canopy response to light can be integrated throughout the day as the solar elevation and solar flux change. Day length, β, and azimuth can be computed as a function of time of day, day of year, and latitude using fairly standard equations. Then, using a full sine-wave function described by Charles-Edwards and Acock (1977), their Eq. [7], the daily-radiation integral can be converted into instantaneous PPFD at any given time of day. This allows the prediction of instantaneous canopy assimilation throughout the day and subsequent integration throughout the day. The diurnal PPFD curve in Fig. 7–4 was generated with the full sine-wave function, using input of 45 mol PPFD m^{-2} d^{-1}.

The daily integral of canopy assimilation vs. the daily integral of PPFD is illustrated in Fig. 7–5 for the simple analytical and sunlit-plus-shaded LAI models presented above in Fig. 7–2. When integrated throughout the whole day, photosynthetic differences attributed to the extinction coefficient [$K_d = 0.5/\sin(\beta)$ or $1.0/\sin(\beta)$] are less important than at solar noon, in part because β varies throughout the day, thus increasing the effective pathlength through the canopy. The simple model presented by Sinclair (Ch. 6 in this book) is also shown in Fig. 7–5; it uses $K_d = 0.5/\sin(45°)$ because it assumes a fixed β (45°) at all times. His model predicts somewhat greater daily P_{can} than the sunlit-shaded-LAI version with $K_d = 0.5/\sin(\beta)$. Notice that daily canopy assimilation vs. daily PPFD integral is somewhat more linear than is instantaneous canopy assimilation vs. instantaneous PPFD (Fig. 7–2 vs. 7–5). This is especially true when one considers that daily PPFD during the summer months really only varies between 20 and 50 mol m^{-2} d^{-1}. This illustration concurs with the point made by Sinclair, that canopy assimilation response to PPFD on a daily basis is fairly linear (thus giving fairly constant daily light-use efficiency) across the range of daily PPFD experienced under field conditions.

Using the hedgerow assimilation model, daily assimilation response to daily PPFD (Fig. 7–6) was simulated for canopies differing in LAI and the ratio of canopy width to row spacing (W/R). It is obvious that incomplete canopy closure ($W/R = 0.5$) results in reduced photosynthesis (associated with reduced light interception). Moreover, it is also apparent that the effect of incomplete canopy closure on assimilation is greater at high LAI than at low LAI (LAI 4.5 vs. 1.5 in Fig. 7–6).

The hedgerow assimilation model response to LAI is presented in Fig. 7–7 for a closed canopy vs. hedgerow situations with varying degrees of

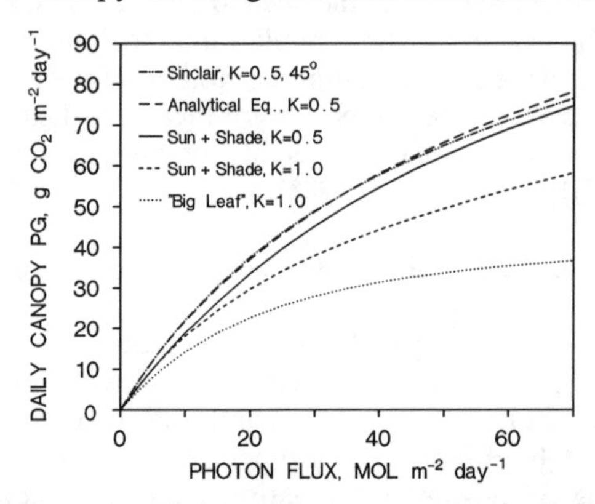

Fig. 7–5. Daily canopy assimilation response to daily PPFD for the same model situations as Fig. 7–2: (a) big-leaf canopy; (b) canopy rate for horizontally angled sunlit-plus-shaded LAI [$K_d = 1/\sin(\beta)$, horizontal leaves, random]; (c) canopy rate for spherically angled sunlit-plus-shaded LAI [$K_d = 0.5/\sin(\beta)$, spherical leaf-angle distribution, random]; (d) canopy rate using the analytical equation of Acock et al. (1978) with $K_d = 0.5/\sin(\beta)$; and (e) canopy rate using sunlit-plus-shaded LAI approach of Sinclair [$K_d = 0.5/\sin(45°)$]. Simulations assume LAI = 5.0, $\sigma = 0.20$, $Q_E = 0.0524$ mol mol^{-1}, and $P_{max} = 1.0$ mg CO_2 m^{-2} s^{-1} for simulations on 1 July at 30° N latitude.

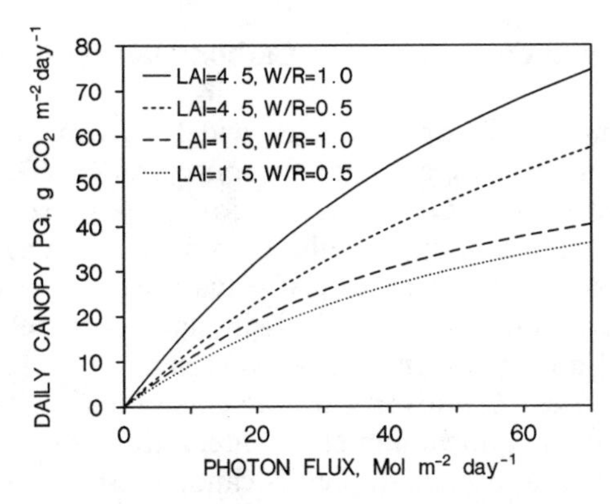

Fig. 7–6. Daily crop assimilation for hedgerow model in response to daily photon flux for canopies differing in LAI (1.5 vs. 4.5) and differing in degree of canopy closure (ratio of canopy width to row spacing, W/R = 0.5 or 1.0). Daily PPFD was 45 mol m^{-2} d^{-1}, σ = 0.20, Q_E = 0.0524 mol mol^{-1}, and P_{max} = 1.0 mg CO_2 m^{-2} s^{-1} for simulations on 1 July at 30° N latitude.

canopy closure, i.e., W/R was varied from 0.2 to 1.0. Daily PPFD was held at 45 mol PPFD m^{-2} d^{-1} and a spherical leaf-angle distribution was assumed. There is considerable interaction between the effect of LAI and canopy closure on daily assimilation. Below LAI of 0.5, incomplete canopy closure has minimal effect on photosynthesis, but as LAI increases, daily canopy assimilation is increasingly limited by incomplete canopy closure. In actual field situations, canopy closure, height, and width tend to increase as LAI increases, thus the limitation is more dependent on final canopy cover relative to row spacing. Assimilation response to the ratio W/R is nonlinear because of diurnal variation in β, and because of interception of diffuse irradiance. In fact, a canopy with 80% cover and an LAI of 6 is predicted to have an assimilation rate only 6% less than that of a closed canopy.

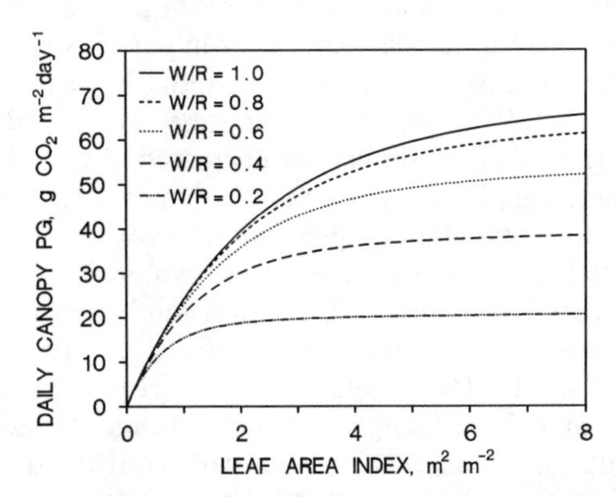

Fig. 7–7. Daily canopy assimilation for hedgerow model in response to LAI for canopies varying in degree of canopy closure (ratio of canopy width to row spacing, W/R from 0.2 to 1.0). Daily PPFD was 45 mol m^{-2} d^{-1}, σ = 0.20, K_d = 0.5/sin(β), Q_E = 0.0524 mol mol^{-1}, and P_{max} = 1.0 mg CO_2 m^{-2} s^{-1} for simulations on 1 July at 30° N latitude.

Canopy Assimilation Response to Carbon Dioxide Concentration

The Farquhar C_i/C_a approach is widely accepted for modeling single-leaf photosynthesis response to CO_2 concentration. This approach for predicting response to CO_2 concentration has more recently been incorporated into some canopy assimilation models, such as the numerical-simulation approaches to integrate leaf equations across canopy depth. (Readers are referred to the numerical-simulation models presented by Gutschick [Ch. 4] and by Norman and Arkebauer [Ch. 5] in this book.) Gutschick (Ch. 4) uses C_i and vertical gradients in leaf and environmental traits, but does not use typical numerical integration. Acock (Ch. 3 in this book) uses a different approach to predict canopy response to ambient CO_2 levels. The Sinclair model (Ch. 6) and the hedgerow version above do not presently consider CO_2 concentration effects.

Canopy Models and Efficiency of Resource Use

Crop models have potential applications for those interested in predicting the efficiency of resource use (efficiency of producing dry matter per unit intercepted solar radiation or water transpired or per unit plant N).

Transpirational Water-Use Efficiency: Improving transpirational water-use efficiency (WUE_t) was an objective of two chapters in this book (Evans & Farquhar, Ch. 1; Gutschick, Ch. 4). At the leaf level, WUE_t is a function of the gas-phase gradient for CO_2 uptake divided by the gradient for water-vapor loss. At a given C_a, WUE_t can be increased by decreasing C_i. Evans and Farquhar, in Ch. 1, describe the theory and the practice of analyzing plant tissue for the ratio of $^{13}C/^{12}C$ as an estimate of the degree to which a given genotype had a lower C_i. A lower C_i is possible either from partial stomatal closure or from increased mesophyll conductance (via thicker leaves with more cells and more rubisco per unit leaf area). In the first possibility, photosynthesis is reduced from partial stomatal closure and lower C_i per se. We suspect that much of the reported gain in WUE_t obtained by the $^{13}C/^{12}C$ technique may be associated with reduced photosynthesis. The other possibility for decreased C_i comes from increased mesophyll conductance (usually associated with more cells and more rubisco per unit leaf area). Assimilation per unit chlorophyll may be less with decreased C_i, but the rate per unit leaf area would be greater. Gutschick (Ch. 4) used a detailed numerical-simulation model with complete energy balance to compare the optimal gain in WUE_t due to lower C_i and/or increased SLM of alfalfa (*Medicago sativa* L.). He even accounted for the negative effect of high SLM on light capture and canopy assimilation during early growth. Gutschick's simulations generally confirmed that increases in WUE_t from lower C_i and higher SLM came at the cost of a somewhat reduced dry-matter yield; the best gain in WUE_t was only 4% if the goal was no decrease in assimilation and biomass yield. When SLM and C_i were both optimized, Gutschick projected maximum gains in WUE_t of 25% with an 11% loss in yield. His is an excellent example of model application to design a crop ideotype in a crop breeding program.

Radiation-Use efficiency: The prediction of radiation-use efficiency (RUE) (dry matter produced per unit intercepted solar radiation) was a major focus of two chapters in this book (Sinclair, Ch. 6; Norman and Arkebauer, Ch. 5). The concept of a relatively constant RUE has great potential for simplifying prediction of plant productivity. Computing RUE requires first the prediction of canopy CO_2 assimilation, and then the adjustment for growth and maintenance respiration to convert to net dry matter gained. In his approach, Sinclair used simple equations to predict daily gross canopy assimilation as a function of daily irradiance, LAI, and leaf N. Then, he allowed for a maintenance-respiration cost of about 15% of daily gross photosynthesis, and a growth-respiration cost of an additional 25 to 35% of daily assimilation, depending on the species. After accounting for maintenance and growth respiration, the net conversion coefficient for soybean, for example, was 0.5 g dry matter g^{-1} glucose. Growth-respiration costs are stable and easily predicted from tissue composition, following the approaches of Penning de Vries and van Laar (1982) or McDermitt and Loomis (1981). By contrast, the magnitude of maintenance respiration is not well known.

The RUE approach used by Norman and Arkebauer is more complex in terms of the model for predicting canopy assimilation. Not only is their model more complex (numerical, layered, energy and mass balance),but they include leaf respiration in their canopy assimilation estimates. To obtain RUE, they separately estimated and subtracted growth and maintenance respiration associated with nonphotosynthetic tissues.

Sinclair argues that RUE is reasonably constant within a given species under good growth conditions, but further states that RUE can be reduced under water and nutrient, particularly N, deficiencies. Although Norman and Arkebauer also liked the concept of a relatively constant RUE, they recommended that RUE coefficients be used with caution in simple models because of the potential for systematic errors.

Optimization Strategies with Canopy Models

There is considerable potential for using canopy assimilation models to ask "what if" questions and to conduct optimization analyses relative to crop traits for particular environments. This is particularly true for cases where canopy models are fully integrated into crop-growth models, to better account for costs and feedbacks. With such models, one can evaluate effects on P_{can} and crop yield attributable to photosynthetic pathway (C_3 vs. C_4, C_i/C_a ratio, specific leaf mass, specific leaf N (SLN is SLM multiplied by N concentration), leaf display, canopy geometry, leaf longevity, and relative durations of vegetative and reproductive growth. Gutschick (Ch. 4 in this book) discusses necessary trade-offs to improve water-use efficiency, whereby gains in water-use efficiency are proposed to be possible with increased SLM and lower C_i/C_a, but at the expense of less dry-matter gain.

Higher leaf P_{max} can be achieved by increasing SLM and SLN; nevertheless, there is a cost to canopy assimilation during early vegetative growth that is readily apparent with a crop-growth simulation. With limited initial C and N, a plant with high SLM and SLN will initially have less leaf area, less light interception, and lower P_{can} during vegetative growth. The P_{max} advantage from higher SLM will take time to be expressed as a growth advantage. As shown by Gutschick and Wiegel (1988), the P_{can} response to SLM depends on the amount of leaf mass and the irradiance level. Optimum SLM is lower when irradiance is low or when leaf mass (LAI) is small. The net effect of a given photosynthetic trait on yield can be analyzed with crop-growth models to account for interactions with length of vegetative period, seasonal weather, and management practices. For example, a crop that has a short vegetative period may gain more benefit from low SLM to improve light interception than it loses because of lower P_{max}. A similar situation may apply for forage crops that are grazed or harvested at relatively low LAI.

There are also potential benefits for P_{can} from optimizing vertical distribution of SLM or SLN within closed crop canopies. The optimum strategy for P_{can} with a closed canopy is to have upper leaves with high SLM and SLN to maximize upper-leaf response to high irradiance, and for SLM and SLN to decrease progressively toward the bottom of the canopy. An optimized vertical distribution of SLN has been predicted to increase daily P_{can} by 3 to 20% compared with a uniform distribution of SLN (constant SLN) (Field, 1983; Hirose & Werger, 1987). The effect is much less for an open canopy. Interestingly, crop canopies already tend toward an optimum distribution of SLN and SLM, so that an optimized distribution of SLN is predicted to improve P_{can} only 5% over the actual SLN distribution (Hirose & Werger, 1987). Similarly, for a crop that has a low average SLM, optimizing vertical distribution of SLM allows a substantial improvement in P_{can} even while maintaining a low average SLM (Gutschick & Wiegel, 1988). This allows high LAI and ability to compete with other species. Gutschick and Wiegel (1988) illustrated that vertical distribution in SLM in crop canopies already tends toward an optimum, thus the potential gain is small. Natural profiles in SLM and SLN appear to be related to vertical profiles in cumulative irradiance receipt, leaf aging, and redistribution of nutrients from old to younger leaves.

The net consequences of single- or multiple-factor changes on grain yield or forage productivity can be evaluated with models that combine canopy assimilation with seasonal growth. One of the features of increased P_{max} is that it is defined at light saturation. While an increase in P_{max} influences assimilation by upper leaves at midday, it has little effect on assimilation by lower shaded leaves or at times when irradiance is low. The advantage of increasing P_{max} dissipates as one goes from instantaneous leaf rates to P_{can} on a daily or seasonal basis. As a result, the increase in dry-matter productivity and radiation-use efficiency becomes asymptotic as P_{max} is increased (see Sinclair, Ch. 6 of this book). One does not get a proportional increase in productivity for each increase in P_{max}. For example, a 10% increase in P_{max} for the canopy assimilation model in the SOYGRO crop model results in only a 4.6% increase in soybean seed yield. Readers will recognize that these strategies have important implications for plant breeding and management, and that they can only be addressed when such canopy models are placed in whole-crop-simulation models. Readers are encouraged to imagine how such integrated canopy–crop models can be used to evaluate strategies to optimize plant traits for best performance under climatic change, drought stress, and stresses from temperature extremes.

Future for Modeling Leaf and Canopy Assimilation

We hope that this collection and review of approaches for modeling leaf and canopy photosynthesis will serve as a resource for present crop scientists and crop modelers, as a focal point for further improvement of models, and as an inspiration for research. Models for single-leaf response to PPFD, CO_2 concentration, and temperature have been succinctly described in the first two chapters (Evans & Farquhar, Ch. 1; Harley & Tenhunen, Ch. 2). Norman and Arkebauer (Ch. 5) and Gutschick (Ch. 4) illustrated how numerical, layered simulation canopy models describing complete radiation, energy, water vapor, and CO_2 balance among leaf strata can be used to predict whole-canopy response to PPFD, CO_2 concentration, wind speed, humidity, and temperature. Gutschick showed how it is possible to reduce the complex numerical model into a summary model to draw some important implications for agricultural production and water-use efficiency. Acock (Ch. 3) introduced alternative approaches for predicting whole-canopy response to PPFD, CO_2 concentration, and temperature. Sinclair's chapter (Ch. 6) contributed an important advance in simple canopy models, particularly to incorporate the effects of leaf N content on canopy assimilation. We have reviewed the approaches taken by the various authors, systematically described single-leaf and canopy assimilation processes, and presented simplified equations for predicting canopy assimilation response to PPFD, LAI, and degree of canopy coverage.

REFERENCES

Acock, B., D.A. Charles-Edwards, D.J. Fitter, D.W. Hand, L.J. Ludwig, J.W. Wilson, and A.C. Withers. 1978. The contribution of leaves from different levels within a tomato crop to canopy net photosynthesis: An experimental examination of two canopy models. J. Exp. Bot. 29:815–827.

Allen, L.H., Jr. 1974. Model of light penetration into a wide-row crop. Agron. J. 66:41–47.

Anderson, J.M. 1986. Photoregulation of the composition, function and structure of thylakoid membranes. Ann. Rev. Plant Physiol. 37:93–136.

Ball, J.T., I.E. Woodrow, and J.A. Berry. 1987. A model predicting stomatal conductance and its contribution to the control of photosynthesis under different environmental conditions. p. 221–224. *In* J. Biggins (ed.) Progress in photosynthesis research. Vol. 4. Martinus Nijhoff Publ., The Hague.

Björkman, O., and B. Demmig. 1987. Photon yield of O_2 evolution and chlorophyll flourescence characteristics at 77 K among vascular plants of diverse origins. Planta 170:489–504.

Boote, K.J., G. Bourgeois, and J. Goudriaan. 1988. Light interception and photosynthesis of incomplete hedgerow canopies of soybean and peanut. p. 105. *In* Agronomy abstracts. ASA, Madison, WI.

Boote, K.J., and J.W. Jones. 1987. Equations to define canopy photosynthesis from quantum efficiency, maximum leaf rate, light extinction, leaf area index, and photon flux density. p. 415–418. *In* J. Biggins (ed.) Progress in photosynthesis research. Vol. 4. Martinus Nijhoff Publ., The Hague.

Boote, K.J., J.W. Jones, and G. Hoogenboom. 1989. Simulating crop growth and photosynthesis response to row spacing. p. 11. *In* Agronomy Abstracts. ASA, Madison, WI.

Campbell, G.S., and J.M. Norman. 1989. The description and measurement of plant canopy structure. p. 1–19. *In* G. Russell et al. (ed.) Plant canopies: Their growth, form and function. Cambridge Univ. Press, Cambridge, England.

Chapman, H.W., L.S. Gleason, and W.E. Loomis. 1954. The carbon dioxide content of field air. Plant Physiol. 29:500–503.

Charles-Edwards, D.A. 1981. The mathematics of photosynthesis and productivity. Academic Press, London.

Charles-Edwards, D.A., and B. Acock. 1977. Growth response of a chrysanthemum crop to the environment. II. A mathematical analysis relating photosynthesis and growth. Ann. Bot. (London) 41:49–58.

Denison, R.F., and R.S. Loomis. 1989. An integrative physiological model of alfalfa growth and development. Bull. Div. Agric. Nat. Res., Univ. of Calif., Oakland.

de Wit, C.T. 1965. Photosynthesis of leaf canopies. Agric. Res. Rep. no. 663. PUDOC, Wageningen, the Netherlands.

de Wit, C.T. 1978. Simulation of assimilation, respiration and transpiration of crops. Simulation Monogr. PUDOC, Wageningen, the Netherlands.

Duncan, W.G. 1969. Cultural manipulation for higher yields. p. 327–339. *In* J.D. Eastin et al. (ed.) Physiological aspects of crop yield. ASA, Madison, WI.

Duncan, W.G. 1971. Leaf angles, leaf area, and canopy photosynthesis. Crop Sci. 11:482–485.

Duncan, W.G., R.S. Loomis, W.A. Williams, and R. Hanau. 1967. A model for simulating photosynthesis in plant communities. Hilgardia 38:181–205.

Ehleringer, J., and O. Björkman. 1977. Quantum yields for CO_2 uptake in C_3 and C_4 plants: Dependence on temperature, CO_2, and O_2 concentration. Plant Physiol. 59:86–90.

Evans, J.R. 1987. The dependence of quantum yield on wavelength and growth irradiance. Aust. J. Plant Physiol. 14:69–79.

Farquhar, G.D., and S. von Caemmerer. 1982. Modelling of photosynthetic response to environment. p. 549–587. *In* O.L. Lang et al. (ed.) Physiological plant ecology II. New Ser. Vol. 12B. Encyclopedia of plant physiology. Springer-Verlag, Berlin.

Farquhar, G.D., S. von Caemmerer, and J.A. Berry. 1980. A biochemical model of photosynthetic CO_2 assimilation in leaves of C_3 species. Planta 149:78–90.

Field, C. 1983. Allocating leaf nitrogen for the maximization of carbon gain: Leaf age as a control on the allocation program. Oecologia. 56:341–347.

Forrester, J.W. 1961. Industrial dynamics. Massachusetts Inst. Technol. Press, Cambridge, MA.

Gerbaud, A., and M. Andre. 1979. Photosynthesis and photorespiration in whole plants of wheat. Plant Physiol. 64:735–738.

Gerbaud, A., and M. Andre. 1980. Effect of CO_2, O_2, and light on photosynthesis and photorespiration in wheat. Plant Physiol. 66:1032–1036.

Gijzen, H., and J. Goudriaan. 1989. A flexible and explanatory model of light distribution and photosynthesis in row crops. Agric. For. Meteorol. 48:1–20.

Goudriaan, J. 1977. Crop micrometeorology: A simulation study. PUDOC, Wageningen, the Netherlands.

Goudriaan, J. 1982. Potential production processes. p. 98–113. *In* F.W.T. Penning de Vries and H.H. van Laar (ed.) Simulation of plant growth and crop production. PUDOC, Wageningen, the Netherlands.

Gutschick, V.P., and F.W. Wiegel. 1988. Optimizing the canopy photosynthetic rate by patterns of investment in specific leaf mass. Am. Nat. 132:67–86.

Harley, P.C., J.A. Weber, and D.M. Gates. 1985. Interactive effects of light, leaf temperature, CO_2 and O_2 on photosynthesis in soybean. Planta 165:249–263.

Hirose, T., and M.J.A. Werger. 1987. Maximizing daily canopy photosynthesis with respect to the leaf nitrogen allocation pattern in the canopy. Oecologia 72:520–526.

Johnson, I.R., and J.H.M. Thornley. 1984. A model of instantaneous and daily canopy photosynthesis. J. Theor. Biol. 107:531–545.

Kirschbaum, M.U.F., and G.D. Farquhar. 1987. Investigation of the CO_2 dependence of quantum yield and respiration in Eucalyptus pauciflora. Plant Physiol. 83:1032–1036.

Lemon, E. 1969. Gaseous exchange in crop stands. p. 117–142. *In* J.D. Eastin et al. (ed.) Physiological aspects of crop yield. ASA, Madison, WI.

Loomis, R.S., and W.A. Williams. 1969. Productivity and the morphology of crop stands: Patterns with leaves. p. 27–51. *In* J.D. Eastin et al. (ed.) Physiological aspects of crop yield. ASA, Madison, WI.

Machler, F., and J. Nosberger. 1984. Influence of inorganic phosphate on photosynthesis of wheat chloroplasts. II. Ribulose bisphosphate carboxylase activity. J. Exp. Bot. 35:488–494.

Marshall, B., and P.V. Biscoe. 1980. A model for C_3 leaves describing the dependence of net photosynthesis on irradiance. J. Exp. Bot. 31:29–39.

McDermitt, D.K., and R.S. Loomis. 1981. Elemental composition of biomass and its relation to energy content, growth efficiency and growth yield. Ann. Bot. (London) 48:275–290.

Meister, H.P., M.M. Caldwell, J.D. Tenhunen, and O.L. Lange. 1987. Ecological implications of sun/shade-leaf differentiation in sclerophyllous canopies: Assessment by canopy modeling. p. 401–411. *In* J.D. Tenhunen et al. (ed.) Plant response to stress. NATO ASI Ser. Springer-Verlag, Berlin.

Monsi, M., and T. Saeki. 1953. Über den Lickhfaktor in den Pflanzengesellschaften und seine Bedeutung für die Stoffproduktion. Jpn. J. Bot. 14:22–52.

Müller, J. 1986. Ecophysiological characterization of CO_2 exchange in leaves of winter wheat (*Triticum aestivum* L.) I. Model. Photosynthetica 20:454–465.

Ng, E., and R.S. Loomis. 1984. Simulation of growth and yield of the potato crop. Simulation Monogr. PUDOC, Wageningen, the Netherlands.

Norman, J.M. 1979. Modeling the complete crop canopy. p. 249–277. *In* B.J. Barfield and J. Gerber (ed.) Modification of the aerial environment of crops. ASAE, St. Joseph, MI.

Norman, J.M. 1982. Simulation of microclimates. p. 65–99. *In* Biometeorology of integrated pest management. Academic Press, New York.

Norman, J.M. 1986. Instrunentation use in a comprehensive description of plant–environment interactions. p. 149–307. *In* W. Gensler (ed.) Advanced agricultural instrumentation. Martinus Nijhoff Publ., The Hague.

Norman, J.M., and G.S. Campbell. 1983. Application of a plant–environment model to problems in irrigation. p. 155–188. *In* D.I. Hillel (ed.) Advances in irrigation. Academic Press, New York.

Peat, W.E. 1970. Relationships between photosynthesis and light intensity in the tomato. Ann. Bot. (London) 34:319–328.

Peisker, M., I. Ticha, J. Catsky, M. Kase, and H.W. Jank. 1983. Dependence of carbon dioxide compensation concentration on photon fluence rate in French bean leaves and its relation to quantum yield and dark respiration. Photosynthetica 17:344–351.

Penning de Vries, F.W.T., and H.H. van Laar. 1982. Simulation of growth processes and the model BACROS. p. 114–136. Simulation of plant growth and crop production. PUDOC, Wageningen, The Netherlands.

Perchorowicz, J.T., and R.G. Jensen. 1983. Photosynthesis and activation of ribulose bisphosphate carboxylase in wheat seedlings. Plant Physiol. 71:955–960.

Rabinowitch, E.I. 1951. Photosynthesis and related processes. Wiley-Interscience, New York.

Ross, J. 1981. The radiation regime and architecture of plant stands. Tasks for vegetative sciences no. 3. Dr. W. Junk, The Hague.

Sage, R.F., T.D. Sharkey, and J.R. Seemann. 1988. The *in-vivo* response of the ribulose-1,5-bisphosphate carboxylase activation state and the pool sizes of photosynthesis metabolites in elevated CO_2 in *Phaseolus vulgaris* L. Planta 174:407–416.

Sharkey, T.D. 1985. O_2-insensitive photosynthesis in C_3 plants. Its occurrence and a possible explanation. Plant Physiol. 78:71–75.

Sharkey, T.D., J.A. Berry, and R.F. Sage. 1988. Regulation of photosynthetic electron-transport in *Phaseolus vulgaris* L., as determined by room-temperature chlorophyll *a* flourescence. Planta 176:415–424.

Sinclair, T.R., and T. Horie. 1989. Leaf nitrogen, photosynthesis, and crop radiation use efficiency: A review. Crop Sci. 29:90–98.

Spitters, C.J.T. 1986. Separating the diffuse and direct component of global radiation and its implication for modeling canopy photosynthesis. Part II. Calculation of canopy photosynthesis. Agric. For. Meteorol. 38:231–242.

Streusand, V.J., and A.R. Portis, Jr. 1987. Rubisco activase mediates ATP-dependent activation of ribulose bisphosphate carboxylase. Plant Physiol. 85:152–154.

Taylor, S.E., and N. Terry. 1984. Limiting factors in photosynthesis. V. Photochemical energy supply colimits photosynthesis at low values of intercellular CO_2 concentration. Plant Physiol. 75:82–86.

Taylor, S.E., and N. Terry. 1986. Variation in photosynthetic electron transport capacity *in vivo* and its effects on the light modulation of ribulose bisphosphate carboxylase. Photosynth. Res. 8:249–256.

Tenhunen, J.D., J.A. Weber, C.S. Yocum, and D.M. Gates. 1976. Development of a photosynthesis model with an emphasis on ecological applications. II. Analysis of a data set describing the P_m surface. Oecologia 26:101–109.

von Caemmerer, S., and G.D. Farquhar. 1981. Some relationships between the biochemistry of photosynthesis and the gas exchange of leaves. Planta 153:376–387.

Weber, J.A., J.D. Tenhunen, D.M. Gates, and O.L. Lange. 1987. Effect of photosynthetic photon flux density on carboxylation efficiency. Plant Physiol. 85:109–114.